Science for All Cultures

A collection of articles from NSTA's journals

Compiled by Shelley Johnson Carey
Managing Editor, *The Science Teacher*

National Science Teachers Association
Arlington, Virginia

Acknowledgements

This collection of articles was reviewed by Napoleon A. Bryant,former Director of the NSTA Multicultural Education Division (1990–92) and Professor Emeritus at Xavier University, Cincinnati, Ohio; Barbara Foots, President of the Association for Multicultural Science Education (1993–95) and Director of Science, Houston Independent School District, Houston, Texas; Maria Lopez Freeman, Director of the NSTA Multicultural Education Division (1993–95) and Director for Equity, California Science Project, University of California, Office of the President; and Olivia Swinton, Science Teacher, Patricia R. Harris Education Center, Washington, DC. Production assistance was provided by Shirley Watt Ireton, Managing Editor, NSTA Special Publications and Gregg Sekscienski, Associate Editor, NSTA Special Publications.

Produced by Special Publications
National Science Teachers Association
1840 Wilson Boulevard
Arlington, VA 22201–3000

Stock Number #PB–109
ISBN # 0–87355–122–2

Table of Contents

Introduction

School desegregation, affirmative action, and cultural diversity can be perceived as necessary trail stops along the destination to multicultural education. The process has not been an easy one! However, in the United States we have begun, when thinking about science and technological achievements, to move away from an ideology of white, Eurocentric superiority to one that embraces and values the knowledge that people of all races, creeds, and ethnicities have made significant contributions to the quality of scientific and technological life we know today.

What is multicultural science education? Originally the concept was a euphemism used when referring to educating African American, Hispanic American, Asian American, and Native American children. Today, much like the evolutionary development seen in moving from school desegregation to multicultural classrooms, the initial perception of what multicultural science education is has been expanded to encompass teaching children of all races, creeds, and ethnicities and to acknowledge all contributions made by persons in these groups to the areas of science, technology, medicine, engineering, and mathematics; in short, to advancement of science and technology throughout the world. Further dimensions of multicultural science education are the belief that all children can learn and the recognition that all students learn best when what is to be learned is related to life experiences found within the individual's culture.

Throughout the history of human beings, there have been several pivotal points that thrust us into the next century or millennium. The invention of the wheel, making and use of fire, synthetic fibers, gun powder, sailing vessels, antibiotics, atomic and nuclear power, and blood plasma are but a few examples. Multicultural education is the pivotal point in education in the 1990s thrusting us into the 21st century and the second millennium. The concept and its implementation in the United States, and globally, have the potential to establish viable bridges between individuals of all races, creeds, and ethnicities.

Multicultural science education provides equal educational opportunities for all students to learn and enjoy science. These opportunities are predicated upon a respect for the customs, mores, and

Statements of Importance from Articles

Statement	Article	Author(s)
Education that legitimizes the cultural norms of only one culture within a pluralistic society robs students from other cultural backgrounds of self-esteem and contributes to discrimination.	Multicultural Science Education	Atwater
Limited proficiency in English does not preclude a student's ability to learn science; on the contrary, most students can and do learn and comprehend a great deal of science content if it is presented to them in their language.	Equal Opportunity Science:	Mason and Barba
Just as it is becoming more evident that Columbus did not "discover" America, our students need to know that there have been people from all cultures who have contributed to our knowledge of science.	Who Really Discovered Aspirin?	Barba, Ooka Pang, and Tran
...[W]e need to nurture every child's science potential because neither we nor they can afford to lose their future contributions.	Are You Turning Female and Minority Students Away from Science?	Blake
The prediction that by the year 2000 at least 70 percent of Black males will be either unemployed or incarcerated has a high probability of coming to pass if the rate at which [we are] losing Black males is not halted.	Sons, Daughters, Where Are Your Books?	Bryant
It is important that parents of black children see, along with their children, the relevance of science to every day life. The more knowledgeable parents are about its relevance and the more insistent they are, the more likely their children will become interested in science.	Black Women in Science: Implications for Improved Participation	Clark
...[I]nitiatives like the Myerhoff Scholarship Program...can lead to increased numbers of minority students who successfully pursue careers in science and engineering.	Recruiting and Retaining Talented African-American Males in College Science and Engineering	Hrabowski, III and Pearson, Jr.
It is unfortunate that the extraordinary successes some [Asian-American students] are enjoying in a few areas have brushed aside any discussion of their unique problems and have detracted from a need to accurately access the costs and sacrifices involved for those who manage to reach such high levels of achievement.	Asian-American Students: The Myth of a Model Minority	Shih

beliefs applicable to the students' respective cultures. Multicultural science education offers an instructional process characterized by the interrelatedness and sensitivity of the school, its community, teachers, and selected curricula to the students, their parents, home, and cultural environments.

The articles comprising this publication are authored by individuals of culturally diverse backgrounds. The articles present several cogent points (see tables above) that collectively provide the reader with a basic understanding of multicultural science education, its scope, implications for teacher education, individual and national well being, and suggestions for using such an approach as an instructional process.

Statements of Importance from Articles (cont.)

Statements	Article	Author(s)
...[T]o increase the participation of Hispanic-American students in science, public school science curriculum [should be] organized around the science of everyday experience.	Underrepresentation of Hispanic Americans in Science	Rakow and Bermudez
The native American child typically falls further and further behind national norms as he or she progresses through school. This phenomenon has been designated as "progressive retardation."	The Need for Strengthening Native American Science and Mathematics Education	Allen and Seumptewa
...[A]lternative assessment strategies will do little good unless educators implement curricular instructional practices that are also multiethnic and multiracial.	Culturally Relevant Alternative Assessment	Tippins and Fitchman Dana
If we define science and technology broadly as ways of observing, describing, explaining, predicting, and controlling events in the natural world, then it is obvious that each culture has its own characteristic science and technology.	Science Across Cultures, Part I	Selin
Asian science and technology was theoretically sophisticated and technologically advanced, and led to the Scientific Revolution of the West.	Science Across Cultures, Part II	Selin
It is important to teach students that the pursuit of scientific advancement is not a European or North American phenomenon. Many, if not all, cultures have given us insights and "giants" from whose shoulders we can see further than any other previous generation.	Not all of those giants were European	Reichert

Multicultural science education, if implemented nationally, has the potential to address and correct the phenomenon of underrepresentation of students of color in science, mathematics, and engineering. With implementation of multicultural education, the right to life, liberty, and the pursuit of happiness becomes a goal attainable by all Americans.

Our students, our nation, and our world will be better served as we gain in experiences that enhance implementation of an educational process that respects the cultural diversity and dignity of each student.

Napoleon A. Bryant, Jr., is Professor Emeritus, Education Department, Xavier University, Cincinnati, Ohio, and former director of the Multicultural Science Education division of NSTA.

Multicultural Science Education

Assumptions and alternative views

by Mary M. Atwater

Science teachers must strive to meet the needs of their culturally diverse student populations. In *Tomorrow's Teachers: A Report of the Holmes Group,* a number of educational goals are identified that can be fulfilled through a multicultural approach to science education. According to the Holmes Group, teachers should:

1. Acquire knowledge and skill that social scientists and practitioners have applied to the study of children's learning;
2. Present appropriate lessons for particular students and use indirect, but powerful teaching strategies such as role playing and cooperation strategies to increase teachers' instructional effectiveness with diverse groups of at-risk students in the classroom;
3. Eliminate school and teacher stereotypes and expectations that can narrow student opportunities for learning and displaying competence; and
4. Create and sustain a communal setting respectful of individual differences and group membership, where learning is valued, engagement is nurtured, and interests are encouraged.

Another report, *A Nation Prepared: Teachers for the 21st Century,* put out by the Carnegie Forum, equates "minority" students with disadvantaged students. The assumption is that being non-White is in itself a disadvantage. The report also ignores the growing number of "new poor" among non-minority families. Both the Holmes and Carnegie reports presume that the only things preventing students of color from becoming part of our internationally competitive work force are financial support and remedial education.

While professing concerns for these students, the reports do not ask what values and goals different cultures might bring to science education. In addition, the reports make few if any recommen-

FIGURE 1. Identifying discrimination in curriculum materials.

LINGUISTICS—Does the curriculum material use masculine terms such as caveman, forefathers, and mankind? Titles such as mailman, policeman, or businessman? Is the pronoun "he" used to refer to all people? Are women referred to only in relation to men? For example, Mr. Henry Jones, his wife, and children attended the party.

STEREOTYPING—Are boys portrayed as ingenious, creative, brave, athletic, curious achievers, while girls are depicted as dependent, passive, fearful, docile victims? Do Anglo-American males hold the majority of the career-oriented positions? What are their careers? Soldier, farmer, doctor, or police officer? In what roles are males of color shown? Worker, farmer, warrior, Indian chief, or hunter? What of the women? Anglo roles usually include a mother, teacher, author, and princess. Those of color also include mother and teacher, as well as slave, worker, porter, and artist.

INVISIBILITY—How often are women depicted in illustrations? Are the contributions of women and people of color mentioned? Does their visibility decrease as the grade level increases?

IMBALANCE—From whose perspective is the text written? Are western terms used to describe other peoples and groups? Indians instead of Native Americans, tribes instead of nations? Is the perspective slanted? Is more than one point of view ever explored? Does the presentation provide the proper perspective on the contributions, struggles, and contributions of women and other underrepresented groups?

UNREALITY—Does the text deal with controversial topics such as discrimination and prejudice? Does it accurately reflect the real world? For example, are any divorced adults or single-parent families depicted? Or is the typical American family described as two adults, two children, a dog, and a house in the suburbs?

FRAGMENTATION—Are contributions of women and people of color treated as unique occurrences (highlighted in a separate box or chapter), or are they integrated into the main body of the text? Does this type of presentation remove their contributions from the mainstream of history, writing them off as an interesting diversion?

Adapted with permission from: Banks, J.A., and C.A. McGee Banks, eds. 1989. *Multicultural Education*. Boston: Allyn and Bacon.

dations for educating culturally diverse student populations in American science classrooms.

MULTICULTURAL SCIENCE EDUCATION

The authors of the aforementioned reports and other researchers have attempted to define the knowledge base of teaching. One such definition includes the following eight categories for science teachers:

1. General/liberal education including basic skills of reading, mathematics, writing, and reasoning;
2. Science knowledge in the domains in which teaching will occur;
3. Pedagogical science knowledge;
4. General knowledge of pedagogical principles and practice;
5. Curricular knowledge;
6. Understanding of student diversity and individual differences;
7. Performance skills including voice, manner, and poise;
8. Foundations of professional understanding including history and policy; philosophy and psychology; cultural and cross-cultural factors; and professional ethics.

As defined, this knowledge base includes both "what to teach" and "how to teach it". Multicultural education enables us to view science teacher education from another perspective.

There are many definitions of multicultural education. One of the most comprehensive definition describes multicultural education as at least three things: an idea or concept, an educational reform movement, and a process. Multicultural education incorporates the idea that all students—regardless of their gender and social class, and their ethnic or cultural characteristics—should have an equal opportunity to learn in school.

Multicultural science education is a field of inquiry with constructs, method-

REFERENCES

Atwater, M.M. 1986. We are leaving our minority students behind. *The Science Teacher* 53(5):54–58.

Atwater, M.M. 1989. Including multicultural education in science education: definitions, competencies, and activities. *The Journal of Science Teacher Education* 1(1):17–20.

Atwater, M.M. 1990. Multicultural science education: definitions and research agendas for the 1990's and beyond. A paper presented at the annual meeting of the National Association for Research in Science Teaching. Atlanta, Ga.

Banks, J.A. 1981. *Multiethnic Education: Theory and Practice.* Boston: Allyn and Bacon, Inc.

Banks, J.A. 1987. *Teaching Strategies for Ethnic Studies.* Boston: Allyn and Bacon, Inc.

Baptiste, H.P., and M.L. Baptiste. 1979. *Developing the Multicultural Process in Classroom Instruction: Competencies for Teachers.* Lantham: University Press of America.

Brown, J.L., and H.H. Pizer. 1987. *Living Hungry in America.* New York: Macmillan.

The Carnegie Forum. 1986. *A nation prepared: teachers for the 21st century. The report of the task force on teaching as a profession.*

The Holmes Group. 1986. *Tomorrow's teachers: a report of the Holmes groups.* East Lansing, Mich.: The Holmes Group.

Shulman, L.S., and G. Sykes. 1986. *A National Board for Teaching In Search of a Bold Standard: A Report for the Task Force on Teaching as a Profession. New* York: Carnegie Corporation.

ologies, and processes aimed at providing equitable opportunities for all students to learn quality science in schools, colleges, and universities. Therefore, the basic premises in multicultural science education incorporate the following:

1. All students can learn science;
2. Every students is worthwhile to have in the science classroom; and
3. Cultural diversity is appreciated in science classrooms because it enhances rather than detracts from the richness and effectiveness of science learning.

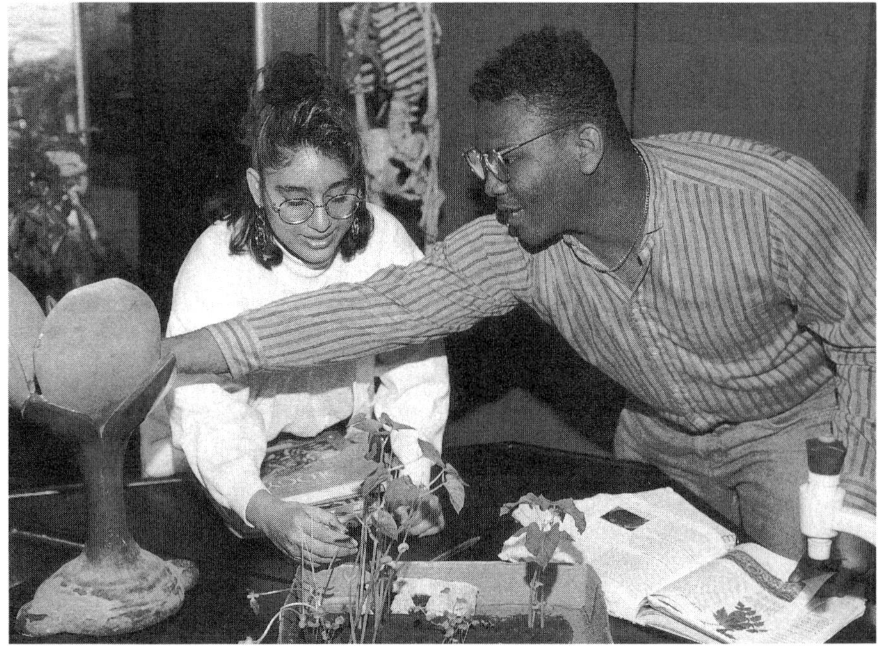
PHOTO BY BRENDA HAMILTON

Thus, advocates of multicultural science education must convince school personnel, parents, and stake holders that science classrooms are now organized so that certain students possess a better chance of being successful in learning science than other students because of their cultural or economic backgrounds.

The school and classroom climates must be altered so all students can be successful in learning science. To institute reform, we must first change the attitudes of our science teachers and other school personnel.

According to some educators, science teachers should possess a profound understanding of the motivations, aspirations, learning modes, linguistics, and culture of their students if they are to be effective science teachers. Students should not be perceived as deficient or disadvantaged simply because they do not share the teacher's beliefs and attitudes. Instead, teachers should try to take advantage of the different cultural perspectives and viewpoints that exist in their classrooms. If we keep an open mind, a multicultural classroom can provide new insights and widen our knowledge base.

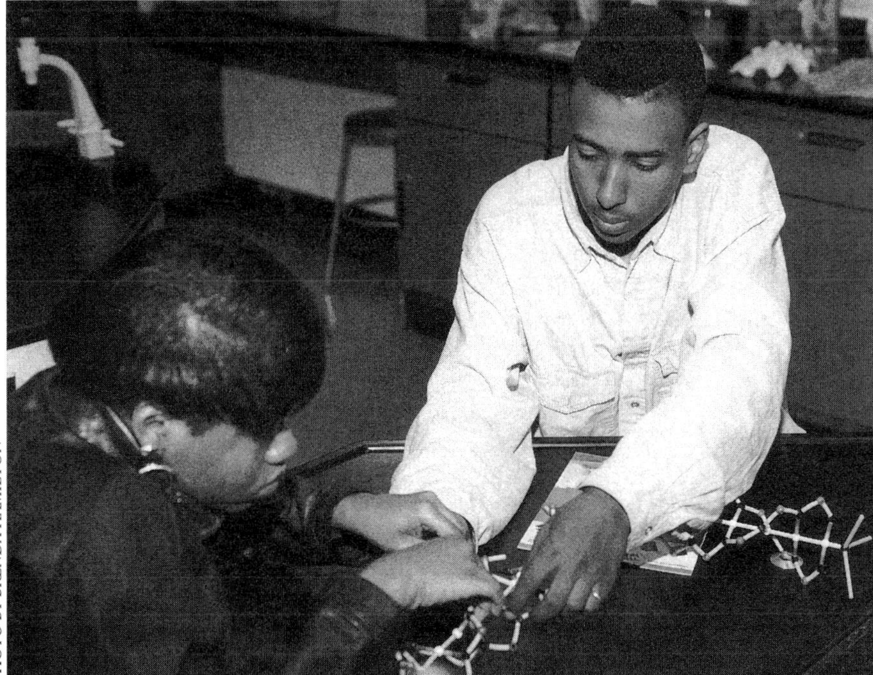
PHOTO BY BRENDA HAMILTON

Science teachers are powerful forces in classrooms. When they hold stereotypes and prejudices, then discrimination can occur. It is well documented that many teachers interact differently with students. Male students are often given extended teacher help to answer questions; females who give wrong answers are usually not asked to elaborate on their answers. Females are often rewarded for the neatness, but not the correctness, of their work. African American males are ignored in science classrooms until they misbehave. Few African American females are praised for their science learning. Native Americans are often introduced to science concepts in a manner that is not meaningful to them.

Education that legitimizes the cultural norms of only one culture within a pluralistic society robs students from other cultural backgrounds of self-esteem and contributes to discrimination. Consequently, science teachers who are willing to take a multicultural approach provide opportunities for their students to learn that many cultures have made contributions to the discipline called science. No single culture has had a monopoly on the generation of science knowledge (see Selin, page 38). These same teachers recognize the potential for students from different cultures to contribute to science because of their different ways of understanding and experiencing the world.

FIGURE 2. A self-directed analysis of multicultural teaching.

1. Classroom climate	List efforts to create a positive learning environment for students to feel at ease in class regardless of gender, race, language, culture, or handicap.
2. Creative curricula	List efforts to supplement the standard curricula with culturally diverse information, activities, and materials.
3. Complimentary communication	List efforts to use both oral and written statements to show respect for the ethnic identity of students.
4. Planned progress	Identify the methods employed to assess the ongoing progress of student learning, as well as the methods employed to prevent students from blaming or using their ethnicity as an excuse for underachieving.
5. Particular practices	Identify monocultural teaching practices and skills that have been replaced with multicultural practices.
6. Productive performance	Identify the multicultural factors that either interfere with teaching performance or restrict productivity.
7. Reflective teaching	Describe the approach for thinking reflectively about unfair situations or inequalities that occur as a result of making certain teaching decisions.
8. Reliable tools	Describe the teaching tools relied on most in helping students understand and accept the cultural diversity of others, such as electronic devices, audio-visual aids, print media, and photographs.
9. Relevant training	Describe involvement in professional development or training activities to enhance knowledge about multicultural education.

Adapted with permission from: Bey, T.M. 1993. *Does Successful Teaching Include Multicultural Goals, Skills and Approaches?* St. Cloud, Minn.: St. Cloud State University.

Multicultural science teachers are knowledgeable about the cultures of their students. How do teachers acquire this knowledge? Reading about different cultures and meaningful interactions with people unlike themselves provide science teachers with this knowledge base. Preservice science teachers must become biethnic, that is, these preservice science teachers must be able to participate successfully in their own ethnic culture as well as another ethnic culture. It is imperative that science teachers possess biethnicity because they must work with the students, parents, and other members of the ethnic community.

Various teaching strategies as recommended in the Holmes report are used by multicultural science teachers so that their students can be successful in learning science. These science teachers understand that a single cultural approach to science continues to alienate groups of students. In order to determine appropriate teaching strategies for students, science teachers must be familiar with the literature about learning styles, perceptual styles, and cognitive styles. It has been documented that most science teaching is geared toward the analytical, field-independent student. Science teachers must possess a repertoire of teaching strategies for relational, field-dependent students as well.

Most students are able to learn what they wish to learn. If students can learn the intricate lyrics of rap songs, then they should surely be able to learn science. The trick is to make science relevant to the student's world. Teachers should also understand that different cultures promote different behaviors. Some tend to reward cooperation while others attempt to instill competitiveness in their young. Students with a competitive spirit may need to be taught how to cooperate with others, and vice versa.

Even if a group of students seem homogeneous, teachers must seek out curriculum materials that respect and relate to all cultures and ethnic groups. Linguistic and cultural biases not only must be eliminated from science teachers, but also from science curriculum materials and assessment. Unfortunately, many multicultural science teachers must develop their own science curriculum materials or purchase additional science posters and books to supplement science textbooks. A tool (Figure 1) has been developed to help evaluate the treatment of women and minorities in curriculum materials.

Science assessment tools include state mandated standardized tests, textbook tests, and teacher-made tests. Little consideration has been given to whether these tests are biased. Few science teachers think about using assessment techniques instead of paper-and-pencil tests. Nonetheless, if science teachers believe students learn differently, then different methods must be used to assess their knowledge and skills.

IMPLICATIONS FOR EDUCATIONAL REFORM

In most cultures, knowledge is socially distributed. It is no different in the United States. However, one of the goals of science education reform is the redistribution of science knowledge to other groups in the United States, besides White, middle-class males. In order to accomplish this goal, we must examine what science knowledge is distributed, what vehicles are used for distributing this knowledge, and who is involved in generating this knowledge.

Many policy makers believe that for the United States to continue to be a leader in science, it is necessary that many others be involved in this endeavor. Therefore, the concept of the pipeline has been developed. To involve more and different people in science, more and different people must enter the educational pipeline. Unfortunately, multicultural science educators suspect the pipeline itself. A new pipeline, one that will celebrate diversity, needs to be constructed.

In the past, the system tried to force all students into the White, middle-class, male mold. Fortunately, not everyone wanted to "standardize" their students, by adopting a definition of an ideal student that was not their own.

Fortunately, not everyone wanted to "standardize" their students by adopting a definition of an ideal student that was not their own.

It is time for students to experience something different in the science classroom. Guidelines are available to help teachers and school personnel eliminate sexism, racism, and discrimination. *Developing the Multicultural Process in Classroom Instruction,* by H.P. Baptiste, Jr., and M.L. Baptiste (University Press of America), includes a very useful checklist for racism and sexism. This tool deals with the community, school board, administration, teachers, guidance staff, and the students are included in this checklist. Figure 2 is yet another checklist for evaluating multiethnic education programs.

Changes must occur in science teachers, school administrators, the school environment, and the science classroom. Otherwise, many Americans, particularly students of color, will be left behind in our science classes.

Mary M. Atwater is a professor in the Science Education Department of the College of Education at The University of Georgia, 212 Aderhold Hall, Athens, GA 30602–7126.

PHOTO BY BRENDA HAMILTON

Equal Opportunity SCIENCE

We need to adapt if students are to achieve

ART BY CHARLES BEYL

y Cheryl L. Mason
nd Robertta H. Barba

In its 1991 position statement on Multicultural Science Education, NSTA stated that teaching science to culturally/ethnically diverse learners is a high priority item for science teaching in he U.S. The statement points out that cience teachers have a responsibility o provide diverse learners with information concerning career opportunities n science and science-related fields.

The bilingual/bicultural, Hispanic/ atino community is the fastest growng and poorest minority group in this ation. Projected figures indicate that Hispanics will soon be the majority opulation is many major cities. Unortunately, this group is also the most lisenfranchised from science, mathenatics, and other related technological areers. Literally millions of Hispanic/ atino students are excluded from this nation's future job market because they lack training in science, mathematics, and engineering.

As a nation, we need to expand the pool of qualified members of the scientific and technological work force by graduating scientifically and technologically capable students from all ethnic backgrounds. It is time for science teachers to reach beyond the differences in languages and cultures and seek ways to turn more students on to science. Science *should* be for all Americans.

CONTEXT OF SCIENCE EDUCATION

While many bilingual programs have been established, such as English as a Second Language, Sheltered English, Transitional English, and the Immersion Model, bilingual/bicultural students rarely receive science instruction at either their appropriate grade level or in their primary instructional language. Presently, approximately 13 percent of the children in California schools are bilingual/bicultural students, a situation soon to be a reality in many states. What must school systems do in order to address the needs of these students?

Previous research concerning limited English proficiency (LEP) students has shown that science instruction is frequently characterized as:

- Being taught by bilingual classroom aides rather than certified science teachers who tend to be monolingual English speakers;
- Relying heavily on the use of worksheets, drill and practice, and other activities that focus on low-level thinking skills;
- Occurring in classrooms that lack materials to conduct hands-on instruction;
- Having textbooks that do not correspond to the district's testing program;
- Projecting low expectations for students' science achievement; and,
- Lacking community and business support for science education activities.

The conclusion that schools lack the resources and personnel to meet the educational needs of bilingual/bicultural science students was reaffirmed during a recent study of classrooms in Southern California. The critical shortage of trained bilingual science teachers, a compounding factor, culminates in a situation in which 72 percent of science classrooms dedicated to LEP students are actually taught by monolingual, English-speaking teachers.

As a result, these science students are not receiving instruction in their primary language.

Even classrooms that strive to be bilingual lack the instructional materials necessary to support the science education needs of students. While some classrooms are equipped with textbooks written in Spanish, the ancillary materials to support verbal science instruction exist only in English. Rarely do schools have laboratory manuals, dictionaries, encyclopedias, atlases, posters, audiovisual materials, computer software, or library books in any language other than English.

Figure 1 graphically compares the percentages and languages of each type of printed material found in the recent study.

CONTENT OF SCIENCE EDUCATION

Historically, schools have focused on linguistic variables, rather than specific disciplines, as the most important components for educating bilingual/bicultural children. Science teachers often present their LEP students with a list of vocabulary words to be memorized and regurgitated. At the other extreme, many bilingual children are exposed to science in a lecture situation during which the information is presented in English at a level above their English competency. Consequently, bilingual students often fail to understand the scientific concepts. When science is presented at either extreme, as a list of vocabulary words or as a high-powered lecture in English, there is little or no opportunity for students to acquire the skills of inquiry or problem solving. A mastery of science content in the bilingual or bilingual/bicultural classroom

FIGURE 1. Distribution of printed materials in classrooms by language.

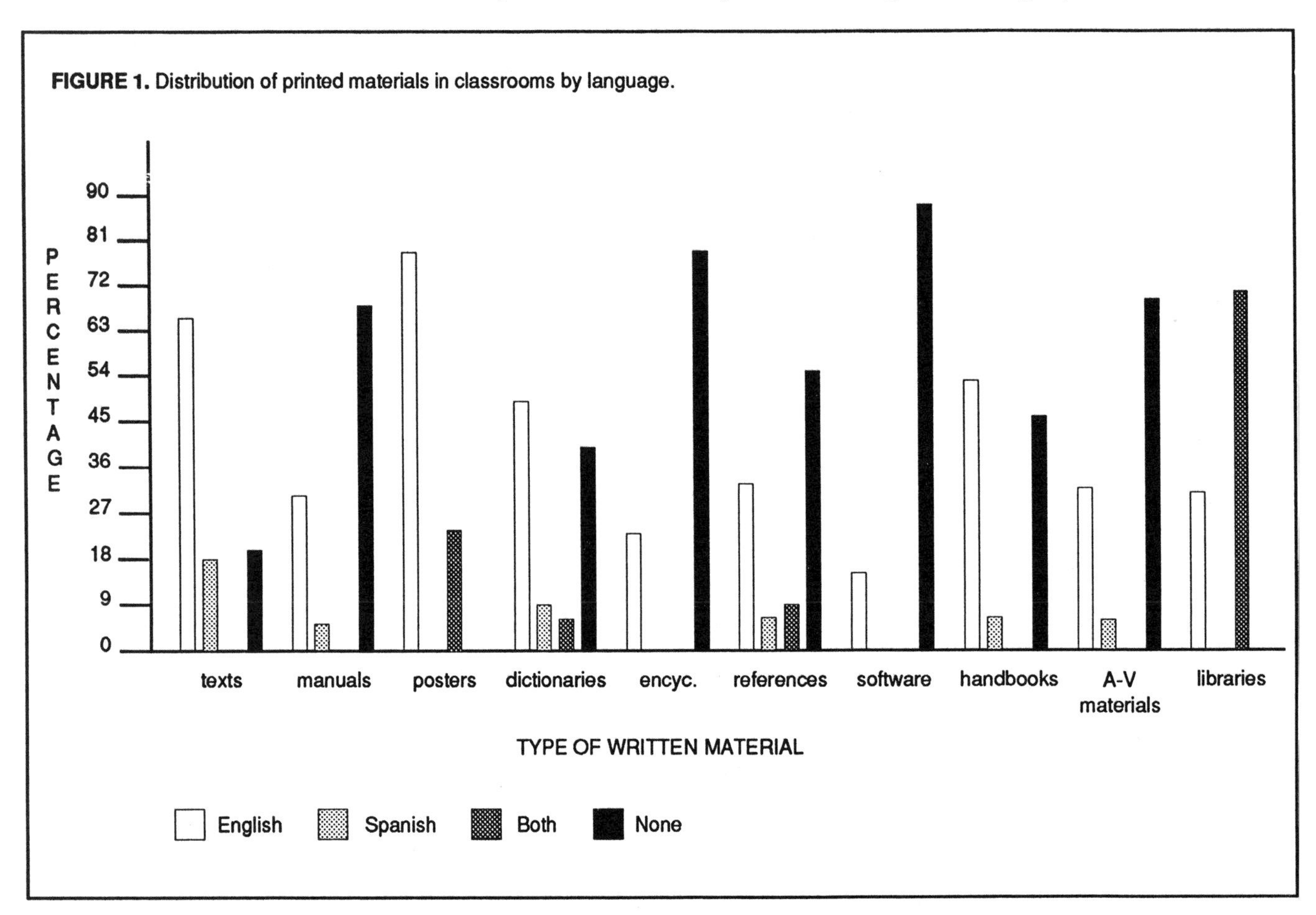

FIGURE 2. Suggested approaches for bilingual/bicultural science teaching.

1. Use hands-on learning whenever possible. Hands-on learning allows students to structure information in their primary language.

2. Use multiple means of representing knowledge. In keeping with Lesh's model of knowledge representation (1983), teachers should present students with the actual object, the spoken and written language (preferably in both languages), models, and line drawings. Multiple means of representing knowledge afford the student the opportunity to form schema in one language or the other or both.

3. Organize cooperative learning groups and peer tutoring. Cooperative groups that match bilingual students with monolingual Spanish speakers afford students the opportunity for peer tutoring. Additionally, teachers should allow peer tutoring to occur in the student's primary language when the students are in small groups.

4. Use culturally familiar analogies whenever possible in order to connect science concepts to the students' real-world experiences. Occasionally bringing culturally familiar examples into the science classroom makes the learning environment hospitable for all students.

5. Supply students with Spanish-language versions of the same textbook being used by their English-speaking classmates.

6. Provide linguistic support materials in Spanish. If supplementary instructional materials are necessary for the learning of monolingual English-speaking students, those same materials are necessary for the learning of bilingual and monolingual Spanish-speaking students.

requires an interplay between language and concept formation.

Even when examples are presented in class, supposedly for the sake of clarification, they are often meaningless to those coming from culturally diverse backgrounds and experiences. When science is taught in this manner, students are frustrated by the instructional level and find science to be boring. It is no wonder that bilingual students tend not to pursue a science career.

Many teachers think that they cannot maintain a high level of conceptual information in the classroom when teaching bilingual students, so they dilute the science content. Limited proficiency in English does not preclude a student's ability to learn science; on the contrary, most students can and do learn and comprehend a great deal of science content if it is presented to them in their language.

IN SEARCH OF SOLUTIONS

Simply labeling a science classroom as bilingual and assigning bilingual/bicultural students to that room is not a solution. The obvious long-term solution to the science education needs of bilingual students is to provide a cadre of trained bilingual science teachers to address the needs of those students. However, there is an insufficient supply of bilingual teachers in the pipeline preparing for jobs in the science classroom. In situations where trained bilingual science teachers are not available, science teachers should engage in classroom activities that will address the needs of bicultural/bilingual students.

Experience has shown that students with a language barrier become more proficient in an environment with hands-on, minds-on activities. Students should learn science by doing science, rather than by reading about it. Visual aids, such as pictures, models, drawings, and actual objects, should be used to enhance oral or verbal instruction. Displaying scientific concepts along with descriptive pictures in the classroom, or using *realia,* effectively reinforces the presented concepts. A large amount of isolated vocabulary is not essential for understanding scientific processes or for developing an awareness of the interconnectedness of scientific concepts. That is why props, cues, body language, and demonstrations help circumvent the perceived language barrier and turn the learning of science into an engaging and meaningful process.

Writing, reading, and speaking skills can be developed through the use of cooperative learning groups, oral reporting, and active note-taking, all of which assist in information processing. Another way to promote language development is to have students paraphrase what is said and/or what they are supposed to know. This also will help monitor their learning. Too often LEP students will claim to understand, rather than have attention drawn to themselves.

It is important to be especially careful when using idioms and abstract terms, and always model proper language and enunciation.

When we use culturally familiar analogies or activities that use examples from a student's own experiences, it gives culturally diverse students the chance to connect new knowledge with

REFERENCES

Bennett, C.I. 1990. *Comprehensive multicultural education (2nd Ed.).* Boston: Allyn and Bacon.

Kessler, C. and M.E. Quinn. 1980. Bilingualism and science problem-solving ability. *Bilingual Education Paper Series* 4(1):1-30.

Lesh, R.A., T. Post, and M. Behr. 1983. Representations and translations among representations in mathematics learning and problem solving. In R.A. Lesh & M. Landau (Eds.). *Acquisition of Mathematics Concepts and Processes.* New York: Academic Press.

National Science Teachers Association. 1991. NSTA releases position paper on multicultural science education. *NSTA Reports!* October/November:7.

National Science Teachers Association. 1989. *Success with minority students in science: The importance of race and ethnicity factors in the classroom. A Handbook for Educators.* Washington, D.C.: NSTA.

Rodriguez, I. and L.J. Bethel. 1983. An inquiry approach to science/language teaching. *Journal of Research in Science Teaching* 20:291-296.

Sutman, F.X., V.F. Allen, and F. Shoemaker. 1986. *Learning English through science. A guide to collaboration for science teachers, English teachers, and teachers of English as a second language.* Washington, D.C.: NSTA.

prior knowledge. As teachers, we should also explore new ways of teaching culturally relevant science, whether it is through herbal medicine (medicinal chemistry), ethnic art, archeo-astronomy, worldwide agriculture, botany, or the history of science. Too often our role models and examples are drawn from a Eurocentric (white male) view of science. It is important to note how other cultures have used or helped develop the scientific disciplines. Figure 2 provides a summary of suggested approaches to teaching science to bilingual/bicultural students.

CHANGING THE CONSTRUCT OF SCIENCE EDUCATION

Good teaching is good teaching is good teaching.... All students find science exciting and relevant when it is taught as an active rather than a passive process. When students can relate what they are learning to their everyday lives, they feel a sense of ownership to the subject. As science teachers, another important consideration is that we must be aware of the needs of our individual students. These diverse needs should be reflected in our curriculum.

Career and academic counseling are also necessary components of our program if we are to attract members of the underrepresented populations to the scientific and technological workforce. Scientific literacy is a goal for all children. We cannot afford to ignore the instructional needs of a large percentage of our students and allow them to remain disenfranchised from careers in science, mathematics, and technology. As a nation, we can no longer afford to ignore the needs of culturally diverse students in our classrooms.

Cheryl L. Mason and Robertta H. Barba are faculty members in the School of Teacher Education at San Diego State University, San Diego, CA 92182-0139.

Who *Really* Discovered Aspirin?

by Robertta H. Barba, Valerie Ooka Pang, and Myluong T. Tran

A colleague of ours who teaches social studies at a large Western university begins his course each semester by asking the question "Who discovered America?" The usual answers of Columbus, Leif Erikson, and other early European explorers are offered by his students. Then he asks a second question, "Was America ever lost?" From a Native American perspective, the answer, of course, is no. Many of us tend to view U.S. history as if life began when European explorers and settlers arrived on these shores.

Our view of the history of science has also been shaped by strong European roots. Women and those from culturally diverse backgrounds are most often relegated to footnote status in science textbooks (See *TST,* October 1991, "Are Women Out of the Picture?") Though these omissions may not be intentional, the hidden message is that women and culturally diverse individuals were not and are not important scientists. Children of color and females lack successful science role models which discourages them from striving for excellence. White males, meanwhile, can often develop a feeling of unfounded superiority, which further aggravates the problem.

SCIENCE IN EVERYDAY LIFE (ROBERTTA BARBA)

As a child, my mother fed me *manzanilla* (Chamomile tea in English) every time I had a fever. My mother learned to use it from her mother, who learned it from her mother. She passed on to me an oral tradition of healing that has existed for thousands of years. We now know that *manzanilla* is a mild analgesic that contains acetylsalicylic acid. However, encyclopedias and science books credit Charles Gerhardt of Germany with the discovery of aspirin (acetylsalicylic acid) in 1853. Who actually discovered aspirin?

After school one day in Quemado, New Mexico, as I was out in the school yard collecting plants for my biology class to use, Geronimo, the school custodian, asked what I was doing. I explained that I wanted to teach my students to identify the names of common range plants that grew there. Slowly, Geronimo smiled and said that perhaps I "should teach children to enjoy and use the plants of the desert." To Geronimo, simply naming plants was not good learning. His perspective and that of many Native Americans and Hispanics were not represented in our curriculum.

In the weeks that followed, Geronimo became a cherished companion and mentor as I walked through the desert around the school. One thing I learned from him is that for thousands of years Native Americans had been selectively breeding corn. Each year the biggest and the best ears of corn were saved by the harvesters as seed corn for the following year. I had not thought of the contribution of Native Americans to genetics research before our walks. Did Gregor Mendel alone discover genetics research or were there others who walked down that same path before him? Who actually discovered genetics research?

Recently, 30 college graduates were asked, "Who invented the compass?" It was with a real sense of sadness that one author found that no one in the group knew that the ancient Chinese scholar, Shen Kua, had invented the compass circa A.D. 1070.

BARRIERS FOR INDIVIDUALS

There are two major barriers to the inclusion of the discoveries of culturally diverse individuals in our science classes. First, the oral tradition often is not recognized as a valid method for transmitting information in the scientific community, but is demoted to "folk medicine."

The second barrier to the inclusion of culturally diverse individuals in mainstream science education is much more complicated. Many culturally diverse researchers have been plagued by a lack of financial resources, lack of collegiality, and a lack of opportunity to participate in mainstream science activities. For example, George Washington Carver taught only in African-American colleges and his laboratory was furnished with discards collected from the city dump, while the laboratories of his white peers were equipped with endowments from their universities. Carver was forced to work within an industrial arts setting, rather than within a mainstream science laboratory. African-American research scientists like Charles Turner, Percy Julian, and Ernest Just were also excluded from full participation in the American scientific community. They were denied the professional interactions, laboratories, and access to public libraries accorded their white counterparts.

Many scientists only gave weight to scientific discoveries that used the European language and standards as legitimate indicators of merit.

In 1721, Onesimus, an African-American slave, explained to his master Cotton Mather how he had been inoculated against smallpox when he was a child. Everyone in Onesimus's tribe knew that it was common sense to transmit a less virulent form of smallpox to children to prevent them from getting a deadly form of the disease. African parents had taught their children this simple inoculation procedure through a tradition of oral history and learning.

Cotton Mather wrote a letter to Dr. Boylston explaining the smallpox inoculation procedure that his slave Onesimus had described to him. Boylston tried the procedure on his son and two slaves. For his discovery, Boylston was called to London where he was honored in the scientific community by being made a fellow of the Royal Society. Onesimus received no such credit. Who, then, discovered the smallpox vaccination?

There exists also a snobbery about the type of scientific discoveries that are considered to be of worth. Many scientists only gave weight to scientific discoveries that used the European language and standards as legitimate indicators of merit.

Just as it is becoming more evident that Columbus did not "discover" America, our students need to know that there have been people from all cultures who have contributed to our knowledge of science. Some scientific practices have been handed down through oral tradition for hundreds of years and these practices were often shaped by many individuals from specific communities. Because our society is Eurocentric, discoveries are often credited to white, male members of our society. Unfortunately, as one source points out, culturally diverse scientists and women have traditionally worked in isolation, lacked the advanced training afforded to other scholars, had their scholarly works usurped by others or their works relegated to footnotes, and have been excluded from full participation in the scientific community.

Our science curriculum could become more accurate and alive if teachers and their students have better access to information about the contributions of culturally diverse individuals and women.

Robertta H. Barba, Valerie Ooka Pang, and Myluong T. Tran are faculty members in the School of Teacher Education, College of Education, San Diego State University, San Diego, CA 92182.

Are You Turning Female and Minority Students Away from Science?

By Sally Blake

As science educators, we all believe that the leaders of tomorrow—who sit in our classrooms today—should possess a high degree of science literacy. In fact, the entire citizenry must be prepared to vote and act responsibly in our highly technological society. For this reason, we involve our students in hands-on learning and activities that attract them to science and then sustain their interest.

But something's not working, at least it's not working with certain segments of the student population. Female and minority students are not keeping up with their counterparts in science, and we've got to do everything we can to combat that dangerous trend. After all, we need to nurture every child's science potential because neither we nor they can afford to lose their future contributions.

The Gender and Race Gap

Currently, females and minorities continue to be underrepresented in science-related employment (Task Force on Women, Minorities, and the Handicapped in Science and Technology, 1989). Studies reveal that, in 1988, females and minorities represented only 33 percent and 2.6 percent, respectively, of all scientists (National Science Foundation, 1990). Consider those numbers in the light of estimates for the year 2000, when 66 percent of those entering the work force will be women and 30 percent will be minorities (Southern Growth Policies Board, 1988). Clearly, we've got to act now to ensure that *all* of our students have the same opportunities and learning experiences during their years of science education.

Differences in science achievement seem to begin developing at the elementary level.

Early Disparities

The National Assessment of Educational Progress (NAEP) data show substantial disparities in science proficiency among subgroups defined by race/ethnicity and gender. While average proficiency for nine-year-old boys and girls is approximately the same (except in the physical sciences), a performance gap becomes evident at age 13 and increases by age 17. In 1986, the mean score for 13-year-old males was 227.3, compared to 221.3 for 13-year-old females (National Science Foundation, 1990). The mean score for nine-year-old black children was 196.2, while white children averaged 231.9. These data reflect the fact that differences in science achievement seem to begin at the elementary level (Mullis and Jenkins, 1988).

Not Enough Science

A contributing problem may be the lack of emphasis on science during the elementary grades. Then, the time devoted to science is less than half that devoted to reading and mathematics (Mullis and Jenkins, 1988).

Yet the problem is even worse for girls and minorities. Data from the NAEP also indicate that girls have significantly less science experience than

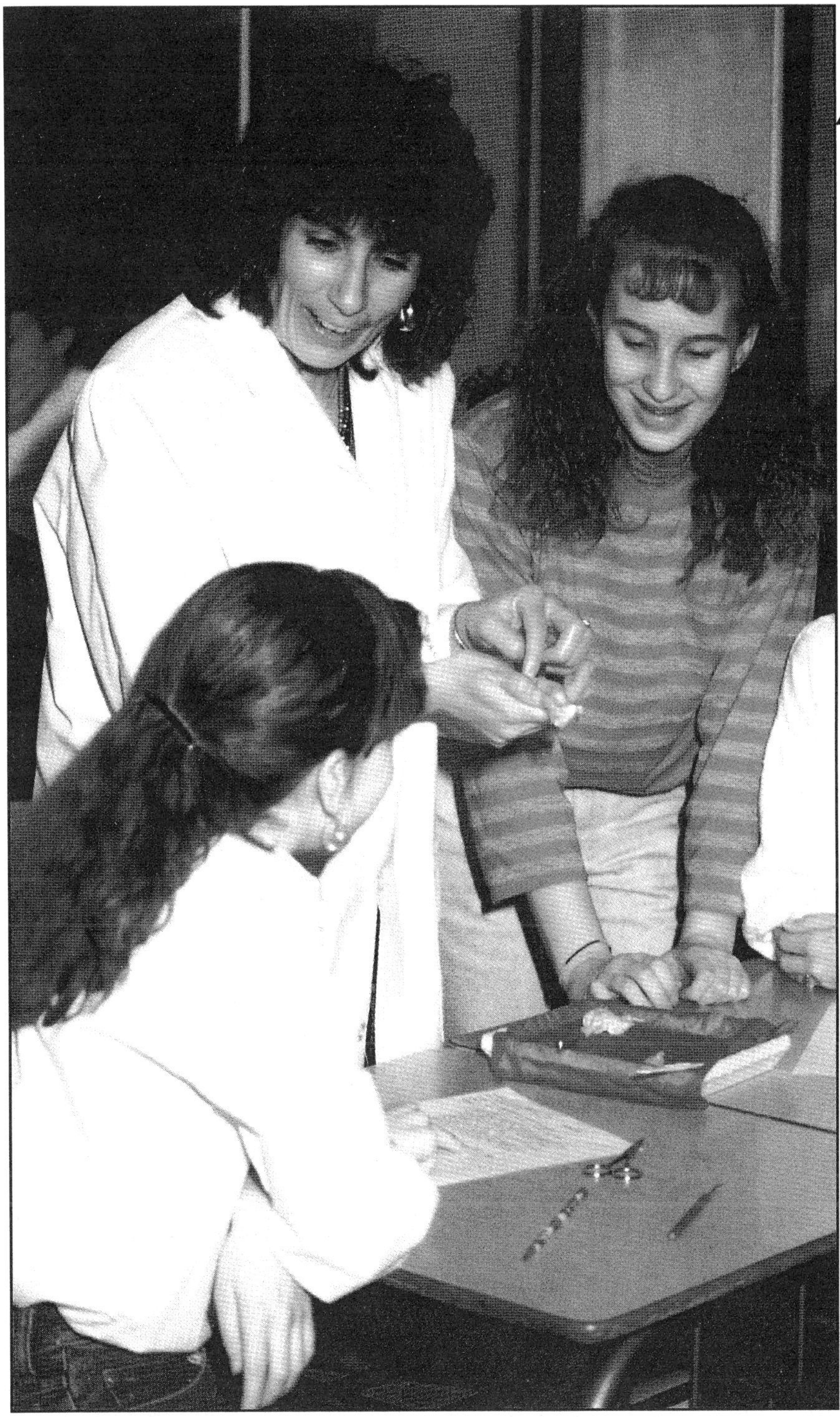

Are these curious students being encouraged to study science?

KATHLEEN M. WABICK

boys at comparable ages. Mullis and Jenkins (1988) note that white students are consistently more likely to report having used various scientific apparatuses than are minorities. Specifically, minorities were less familiar with such simple devices as scales and magnifying glasses. Unfortunately, low-income and minority students are also less likely to have qualified science teachers (National Science Teachers Association, 1990/1991).

"Invisible" Students?

The comparative lack of successful female and minority scientists cannot be traced to any single factor but certainly stems from insufficient academic support, including negative signals that discourage girls and minorities from developing their science abilities (Mullis and Jenkins, 1988). Data collected from programs that attempt to recruit and retain minorities of both sexes suggest that minority females turn away from science for some of the same reasons that white females do. Kahle (1985) and Cambell (1986) have described teachers' instructional effectiveness as significant in attracting girls to science, yet very little research has centered on comparing the relevant teaching methods and approaches.

We must be aware of the link between the overt curriculum, which is observable, and our hidden curriculum, which may affect your teaching more subtly (Sapiro, 1990).

The way that you verbally interact with your students may unconsciously discourage female and minority participation (Rosser, 1990). For example, you may promote the "invisibility" of females by subtle practices, such as calling directly on males but not on females. Studies indicate that teachers do address males by name more often than females, and they also give males more time to answer questions (Association of American Colleges, 1982). Girls, particularly black girls, get less feedback than boys (Irvine, 1986), and minority students often feel ignored or put down by teacher response and instruction (Noonon, 1980).

Addressing Gender Bias

Examining gender bias in the classroom, researchers have isolated several contributing factors. Gilligan (1982) concludes that girls approach problem solving from the perspective of interdependence and relationship rather than from the isolated skill analysis viewpoint favored by boys. Thus, girls feel less comfortable approaching laboratory experiences when they don't understand the relationship of one experiment either to another experiment or to a life experience. If, as research suggests, females learn better in a cooperative, rather than a competitive environment, then scientists should be introduced as individuals wholly integrated with other aspects of daily life (Rosser, 1990).

Baker (1983) contends that there is a conflict between science and the definition of femininity. Other researchers point to a male bias in the choice and presentation of scientific problems (Harris, Silverstein, and Andrews, 1989), and in the design and interpretation of scientific work.

Be Part of the Solution

In order to draw a higher proportion of girls and minorities toward science careers, we must begin by evaluating our own classroom behavior with respect to race and gender discrimination. After that self-examination, consider the following guidelines for structuring science activities that motivate and respond to the interests of *all* students.

The comparative lack of successful female and minority scientists cannot be traced to any single factor, but it certainly stems from insufficient academic support.

1. Choose activities that are free from sexual stereotyping.
2. Spend instructional time on science activities every day.
3. Design activities that will ease the stress of competition.
4. Feature the use of simple science tools in your activities.
5. Emphasize the practical applications of science and how it relates to students' lives.
6. Include a wide variety of science topics and concepts in order to reduce anxiety.
7. Present data on both males and females, whether the subjects are animals or humans, in all laboratory experiences.
8. Give equal feedback to females, males, and minorities when working with science problems.
9. Make a conscious effort to acknowledge the contributions of female and minority students and scientists to scientific observation.

The use of simple science tools in hands-on activities will help to develop fully female and minority students' science potential.

KATHLEEN BROTHERS

It's in Your Hands

In the future, more and more jobs will require a high level of science knowledge. As the demand increases for workers with science skills, any underrepresentation of gender and race subgroups will become more problematic.

Listen to the recommendations of the research done thus far, and call for additional studies that explore these questions in greater depth. Plan science activities that nurture scientific literacy in your female and minority students, so that all your students will be ready to meet the demands of the future.

Resources

American Association for the Advancement of Science. (1990). *Agenda for action (Publication No. 90-13S)*. Washington, DC: Author.

Association of American Colleges. (1982). *The classroom climate: A chilly one for women?* Washington, DC: Project on the Status and Education of Women.

Baker, D. (1983). Can the difference between male and female science majors account for the low number of women at the doctoral level in science? *Journal of College Science Teaching, 13*(2), 102–107.

Bank, B.J., Biddle, B.J., and Good, T.L. (1980). Sex roles, classroom instruction, and reading achievement. *Journal of Educational Psychology, 72,* 119–32.

Cambell, P. (1986, March). What's a nice girl like you doing in a math class? *Phi Delta Kappan,* 516–520.

George, Y.S. (1982). Affirmative action programs that work. In *Women and minorities in science,* American Association for the Advancement of Science Selected Symposia Series. Boulder, CO: Westview.

Gilligan, C. (1982). *In a different voice: psychological theory and women's development.* Cambridge, MA: Harvard University.

Harris, J., Silverstein, J., and Andrews, D. (1989). *Educating the majority.* New York: Macmillan.

Irvine, J.J. Teacher-student interactions: effects of student race, sex, and grade level. *Journal of Educational Psychology, 78,* 14–21.

Kahle, J.B. (1985). *Women in science.* Philadelphia: Falmer.

Mullis, I.V.S., and Jenkins, L.B. (1988). *The science report card: Elements of risk and recovery.* (Report No. 17-S-01). Princeton, NJ: Educational Testing Service.

National Assessment of Educational Progress. (1987). *Science objectives: 1985–86 assessment.* Princeton, NJ: Educational Testing Service.

National Science Foundation. (1990). *Women and minorities in science and engineering.* (Report No. 90-301). Washington, DC: Author.

Noonon, J. (1980). White faculty and black students: examining assumptions and practices. Richmond, VA: Center for Improving Teacher Effectiveness.

Rosser, S.V. (1990). *Female friendly science.* New York: Pergamon.

Sapiro, V. (1990). *Women in American society.* Mountain View, CA: Mayfield.

Southern Growth Policies Board. (1988). *Halfway home and a long way to go.* Little Rock: Bell South.

Staff. (1990–91 December/January). Tracking, poor-quality teaching and resources hinder performance of minority students in science and math. *NSTA Reports,* p. 7.

Task Force on Women, Minorities, and the Handicapped in Science and Technology. (1989). *Changing America: The new face of science and engineering.* Washington, DC: Author.

WILLIAM E. MILLS, MONTGOMERY COUNTY PUBLIC SCHOOLS

With encouragement, all your elementary students can become scientists.

SALLY BLAKE is an assistant professor of education at Kentucky Wesleyan College in Owenboro.

Sons, Daughters, Where Are Your Books?

Positive attitudes about Black students' ability to excel academically, combined with supporting efforts from schools, universities, industry, and private citizens are required if we are to reverse poor academic performance of these students as a group.

Napoleon Bryant

The location is Big City, U.S.A., and secondary school is out for the day. The hundreds of Black students leaving laugh, play, and make plans with each other while climbing into friends' cars or boarding chartered buses or public transit coaches. Some students spill into the streets, narrowly avoiding moving traffic. The time of year is not important. This scene is repeated each day in our major urban areas when black males and females, grades 7–12, leave school for the day. In observing the students leave, a common factor shared by each group becomes evident: few, if any, students have books! What have they done with their books? Don't they need them to prepare for tomorrow's lesson?

But wait—observing dismissal time reveals something else, as well. Some teachers are among the first wave of students to depart the school building. Other teachers emerge shortly after the grounds are clear. Observe carefully and you notice many teachers are without books or other instructional materials. How can this be? Newspapers, magazines, and educators themselves are always drawing attention to the number of classes teachers must teach and for which they must prepare. No opportunity is missed to speak of large class sizes and of the difficulty teachers have in meeting the diverse needs of their students.

We live in a time of rapidly expanding knowledge. Knowledge true today is so modified within five years that the formerly accepted concepts, at best, provide an historical basis for what lies ahead. Sources of knowledge pervade our environment. Telecommunication systems, textbooks, magazines, newspapers, video systems, and the application of scientific knowledge in the production of life-enhancing technology all signal that it is impossible to learn everything in school. Translated: There is not enough time in the short school day and year to learn about everything. Learning must carry over into out-of-school time.

Impact

We have a serious problem. Secondary schools in urban areas increasingly are seen by students and teachers as someplace you go between 8 A.M. and 3 P.M. Interaction with teachers that fosters academic development is minimal. Athletic and other noncognitive programs hold more fascination for the students than the lesson planned for the period. How is it possible with so much to learn, with the need to be scientifically literate, and with the demand for preparation-for-life experiences that the majority of Black stu-

UNIVERSITY OF FLORIDA PHOTO BY WALTER COKER

dents go home daily without books and that the majority of their teachers take the evenings off (as presumed by the absence of instructional materials accompanying them when they leave school for the day). Students will return to school tomorrow unprepared for class. Will their teachers be any different?

What are the implications of the question: Sons, daughters, where are your books? What are the implications when teachers of Black youths relegate formal learning to the time available during the school day, or when they try to accomplish everything between 8 A.M. and 3 P.M.? The social and economic implications are demoralizing in their impact on Black students and the needs of the nation.

The problem Black students have in performing well academically has been well documented. The problem, which continues to be one of the most detrimental facing Black people and the nation today, is further exacerbated by a resurgence of racism throughout American society, our schools, and universities. The American Council on Education (ACE) recently found that Blacks remain the most segregated minority in our nation [2].

Black youths are dropping out of high school in increasing numbers. The prediction that by the year 2000 at least 70 percent of Black males will be either unemployed or incarcerated has a highly probability of coming to pass if the rate at which the classroom is losing Black youths is not halted.

The continuing decline in the number of Black students enrolled at institutions of higher learning signals a return to the period in our history prior to federal and state financial assistance. ACE reports an erosion of gains achieved in minority student enrollment at universities due to the high attrition rate of these students [1]. Boyer presents some very disturbing statistics on absenteeism of Black youths in high schools in major urban communities. On a given day, as many as 40 percent may be absent, and high absenteeism contributes to poor academic performance. Philadelphia and Boston recorded dropout rates of 38 percent and 43 percent, respectively. Even more alarming, 50 percent of students in Chicago failed to graduate in 1984. Boyer further asserts a relationship between poverty and education [4].

Atwater affirms that Black students

are being left behind. She expresses concern regarding the few minority students who are enrolling in science classes in high school or university. An examination of statistics on the national level reveals that minorities received 2.9 percent of the B.S. engineering degrees awarded in 1973 compared to the 5.7 percent in 1981 [3]. But the reported 1981 gain in the number of degrees awarded to minority students in engineering and physical science will erode quickly because fewer minorities are now able to go to college, more minority students are dropping out, and science and engineering tend not to be majors they select.

Universities should recruit academically talented Black students from urban schools with the same aggressiveness they recruit Black athletes.

The Challenge

Learning and excelling academically is reported to be out of vogue with Black students. They do not want to be viewed as "acting white." Secondary teachers cite numerous instances of academically capable Black students capitulating to existing codes of peer group behavior that frown upon the display of "brains" in the classroom. To excel academically is to risk being ostracized by one's peers; to take books home in the afternoon and return the next day prepared for the day's lessons is to risk being adjudged not deserving of group membership.

And teachers are just as subject to these pressures. Very few escape the daily frustrations, bitter disappointment, and agony experienced when they attempt to teach and enhance their students' intellectual development, only to be thwarted in all attempts. The result: teacher capitulation to student peer group behavior and abdication of a major professional responsibility—to promote the growth and development of students to their fullest potential. When the bell rings, students and teachers put their materials away and leave.

Will it be any different tomorrow? How can the trend be reversed so that Black students have equal access to academic, economic, and social prosperity and to fulfillment as worthwhile human beings? If students continue to go home from school empty handed, if teachers capitulate rather than hold tenaciously to high standards of excellence and productivity for their students, America's loss in brain power, as represented by the academically underdeveloped minds of Black students, will be incalculable.

Suggestions

Teachers at all grade levels, kindergarten through 16, must first acknowledge the intrinsic worth and intellectual capabilities of the Black student to achieve at high levels of excellence. Before affixing blame for poor academic performance on the learner, teachers charged with the responsibility of teaching Black students should undertake a systematic and thorough analysis of their teaching styles. We must examine the attitudes and pedagogical skills we use with Black students for their positive impact upon the learner and eliminate any teacher behaviors that inhibit learner success.

Black students are not stupid. They will perform meritoriously in areas from which they derive economic success and personal fulfillment. The accomplishments of Black youths in the fields of sports and entertainment validate their ability to achieve. Black students, from the time they begin formal education, need to understand that it takes as much determination, sacrifice, and hard work to be outstanding in sports and entertainment as it does to pursue a career in science or engineering. The myth that "you have to work and study too hard" to become a scientist or an engineer must be dispelled. Black students need to be shown that development of their minds can bring economic success, personal fulfillment, respect, and appreciation from others. They need to see that the pot of gold lies at the end of the rainbow for those who persevere, who make that mighty reach, and who excel to the extent of their intellectual capabilities.

Society must come to grips with the fact that Black youths can excel at more than sports and entertainment. They need models in the fields of science and engineering as early as elementary school. Other science and engineering personnel from industry and universities could join Black colleagues in establishing working relationships with schools at all levels in urban areas. Such relationships could result in "big scientist–little scientist" twosomes in which Black students would be introduced first-hand to science and engineering fields by Black professionals acting as the students' associates. They would come to see the need to take their books home with them and would be able to get assistance from their associates, if necessary, when preparing for the next day's classes.

Industry and universities can show Black students that it pays to perform well academically by aggressively seeking to hire them at salaries commensurate with their capabilities and experience. Opportunities for promotion for the Black employee should be no different from that of any other employee.

All levels of advancement should be equally accessible to the Black employee. Black role models in science and engineering can help students see the absurdity in believing that to demonstrate academic excellence is to behave like white students. Teachers should work diligently with all students to dispel myths that function to diminish students' inclinations to fully realize their potential. The rebuttal of such myths provides excellent exercises in critical thinking that should help students conclude that to accept performing below their ability levels for the sake of group membership is to have a poor concept of themselves. Their ability to forge a path of success in spite of peer opinions must be strengthened.

Universities, industry, and private citizens can present reasons why Black students should prepare themselves for careers in science and engineering. Employment opportunities must abound and be attractive. In addition, universities should recruit academically talented Black students from urban schools with the same aggressiveness they recruit Black athletes. Further, universities could create full scholarships for Black students who want to major in science or engineering, establish partnership-in-education programs with schools at each level, and provide release time for faculty who develop programs to involve Black students and assist their development and interest in science and engineering.

Programs sponsored by industry to encourage Black students' interest in careers in science and engineering need to be continued and expanded upon. Scholarships established in recent years by certain industries have been successful in providing financial assistance for deserving Black students. These scholarships need to be increased in number and amount. As suggested above, industrial scientists and engineers could work with their university colleagues in establishing big scientist–little scientist pairs. Arrangements could be explored by industry personnel with school systems to grant science credits to students for projects and time experienced under the tutelage of practicing scientists or engineers.

Napoleon Bryant is a professor in the Department of Education, Xavier University, Cincinnati, OH 45207.

Black students will produce if they perceive you are genuinely interested in their welfare. Students in a Brooklyn high school responded positively to a commitment from a philanthropist to pay all college expenses for those who performed well academically and went on to college. More programs of this nature are needed. Other people of wealth may consider underwriting the cost of a university education as a way to address this national dilemma. The precedent has been established.

The challenge of how to get students to take their books home is a real, but not impossible, task. Teachers would welcome reversal of the current trend, but many feel incapable of accomplishing this reversal by themselves.

Black scientists and engineers are needed to begin the reversal. They will not be able to complete the task by themselves. Sensitive colleagues and private citizens, abundant scholarship aid, and aggressive employment practices can do much to highlight reasons Black students should succeed academically.

But first and foremost, Black students themselves must want to succeed and not let anything prevent them from demonstrating their capabilities. □

References

1. American Council on Education. "Minority Student Enrollment Gains Eroded by Attrition." *Higher Education and National Affairs* 36:4; February 23, 1987.
2. American Council on Education. "Videotape Program to Combat Racism." *Higher Education and National Affairs* 36:4; May 4, 1987.
3. Atwater, Mary M. "We Are Leaving Our Minority Students Behind." *The Science Teacher* 53:54–58; May 1986.
4. Boyer, Ernest L. "Early Schooling and the Nation's Future." *Educational Leadership* 44:4–6; March 1987.

Bibliography

American Council on Education. "Study Finds Blacks Remain the Most Segregated Minority." *Higher Education and National Affairs* 36:9; February 23, 1987.

Brown, Ezra. "Wrong Message from Academe." *Time* 129:57–58; April 1987.

Carmichael, J.W. Jr., Sr. Joanne Bauer, and Donald Robinson. "Teaching Problem Solving in General Chemistry at a Minority Institution." *Journal of College Science Teaching* 16:453–457; March/April 1987.

Gorman, Trish. "Schools Where All Our Kids Are Winners." *American Teacher* 71:8–9; October 1986.

Hodges, Helene. "I Know They Can Learn Because I've Taught Them." *Educational Leadership* 44:3; March 1987.

Keeter, Larry. "Minority Students At Risk: An Interview With Professor Shirley Chisolm." *Journal of Developmental Education* 10:18–21; January 1987.

Murphy, Donna M. "Educational Disadvantagement: Associated Factors, Current Interventions, and Implications." *Journal of Negro Education* 55:495–507; Fall 1986.

Riley, Richard W. "Can We Reduce the Risk of Failure?" *Phi Delta Kappan* 68:214–19; November 1986.

Seeley, David. "Education, Dependence, and Poverty." *Education Digest* 52:6–9; January 1987.

Wineberg, Samuel S. "When Good Intentions Aren't Enough." *Phi Delta Kappan* 68:544-45; March 1987.

Black Women in Science

Implications For Improved Participation

As we develop into a more scientific and technological society, we must make a focused effort to illuminate the factors that influence black women to enroll in more science courses and pursue careers in science.

Julia V. Clark

Too few black women are represented among the population of scientists in the United States. Despite substantial gains over the past decade, black women are still underrepresented in science and engineering, both in employment and training [8]. Why should this be in a country that prides itself on providing quality education and equality of opportunities for all of its citizens? Most black women have chosen not to take advantage of this opportunity by not pursuing careers in the traditionally male enclaves of science, technology, and engineering. There is a steady increase in the number of black women in other traditionally male-dominated fields such as business, but very few select a field of science. In view of the importance of science and technology in the United States today, it is crucial that the number of black women pursuing careers in the sciences increases.

Status of Black Women in Science

Although black women, as well as all women, are underrepresented in the sciences, they have made substantial gains both in training and employment since the early to mid-seventies [9]. Of the approximately 512,000 women who were employed as scientists and engineers in 1984, 4.5 percent (23,000) were black. Blacks were more highly represented among women than among men scientists and engineers. Table 1 shows another way of looking at the status of black women in science and engineering fields. While women represent about 13 percent of the total scientists and engineers employed across all racial groups, black women represent 25 percent of all employed black scientists and engineers [9].

At the doctoral level, relatively few of the employed women scientists and engineers were members of minority groups. In 1983, about 3 percent (1,400) were black and 7 percent (3,400) were Asian. Among male scientists and engineers with doctorates, about 1 percent were black and 8 percent were Asian. Black women constituted a larger share of all black doctoral scientists than did other minority women of their respective racial groups. Table 2 breaks down the field distributions for women scientists and engineers [9].

Tenure status and academic rank may also be used as measures of career development. Among doctoral women in educational institutions, blacks are in tenure-track positions slightly more often than white and Asian women. In 1983, about 65 percent of the black doctoral women were in tenure-track positions, compared to approximately 62 percent of white women and 45 percent of Asian women. Among doctoral women, variations in the proportion holding professional rank ranged from 86 percent of Asian women to 89 percent of black women. The number of women receiving postdoctoral appointments in science rose from 900 in 1973 to 3,100 (29 percent) of the appointments in 1983. The appointments for black women increased from 28 to 215 [9].

Regardless of race, salaries for women were below those for men. Salaries for black and Asian women, however, averaged about 78 percent of those for men in these same racial groups, while those for white women averaged about 71 percent of white male salaries. At the doctoral level, salaries for black and white women were higher than those for Asian women—$32,000 for black and whites and about $31,000 for Asians [9].

Barriers to Success in Science

According to NSF, the failure of female and minority students to take the appropriate high school courses is one possible source of continuing disparity [9]. In a survey conducted by NSF in 1982, 47 percent of male high school students had taken 4 math courses and 25 percent had taken 4 science courses, compared with 36 percent in mathematics and 18 percent in science for female students. It was noted, however, that the female students earned higher grades in those courses.

Because of boys' mechanical inclinations, parents tend to encourage boys more than they do girls to pursue scientific careers. In school, girls often receive less encouragement than do boys to take science and math courses. They are seldom guided toward technical fields by teachers and counselors. They are frequently unaware of scientific and technical job opportunities. Another very important factor is that the mass media do little to change the pattern, especially as it pertains to black females.

The perception of the usefulness of science for future education and career plans and the support or lack of support from significant others are among the major factors associated with women's decisions to elect or not to elect to take science courses beyond the required courses for a high school diploma [1]. For black females, these factors in turn are influenced by the stereotype of science as a male domain. Most of the science-related fields, such as engineering, physics, computer science, and the natural sciences tend to be male-dominated and high paying. Other factors associated with course-taking and achievement are attitudes toward science and feelings of self-confidence.

The 1977 assessment of science by the National Assessment of Educational Progress (NAEP) presents data that show that more blacks than whites at age 17 find science seldom or never boring (27 percent compared with 17 percent of whites); 30 percent of blacks find science always or often fun compared with 26 percent of whites; 82 percent of black 17-year-olds compared with 58 percent of whites think that science should be required in school [5]. But these same students report fewer science experiences, find science less useful out of school, and have less confidence than whites in the ability of science to solve current or future problems [3].

Over the last decade, many programs recognized the shortage of both black women and men in science and have tried to do something about it. A multitude of recruitment and intervention programs in science and science related disciplines were developed throughout the United States in an effort to increase the interest and participation of blacks in science and to alter the pattern of precollege education for these students.

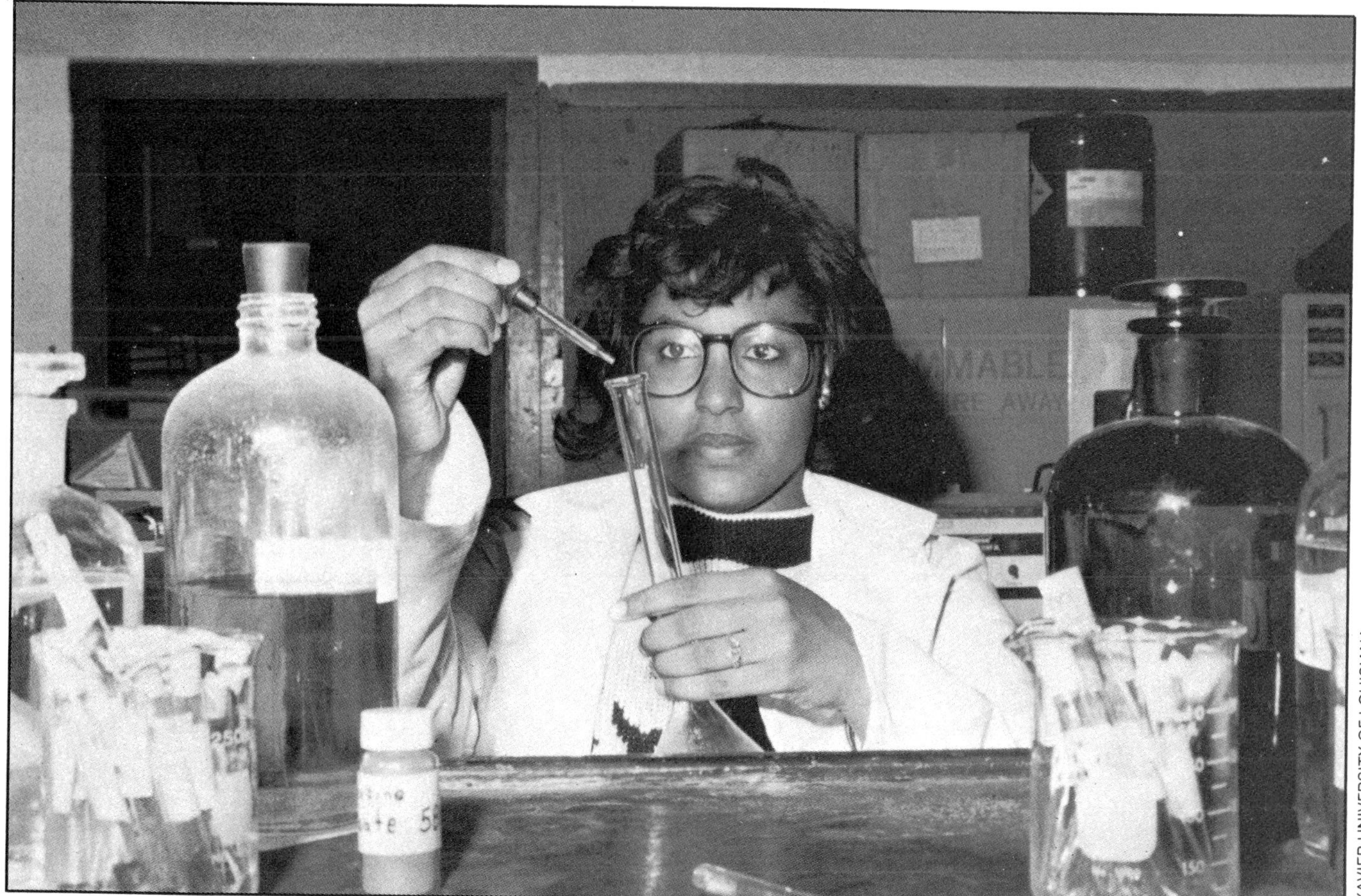

XAVIER UNIVERSITY OF LOUISIANA

Many enrichment and intervention programs are available today, but they are reaching out to the so-called gifted and talented students. Thus, a substantial portion of our population still suffers from the consequences of racial, social, and economic discrimination, compounded by watered standards, poor guidance, and token efforts [2].

A Statement of Concern

One of the major challenges facing education today is how to ensure that all of our students are adequately educated. To meet this challenge, we must mobilize all of our forces and efforts to educate black children for technological survival and maximum economic self-sufficiency.

When considering why black women take fewer science courses in high school than males, why they abandon mathematics at an earlier grade level, or seldom study senior physics or mathematics courses, we generally look for obstacles. Although many have been identified, it could also be that the origin of the problem lies more in what is absent from the black female's education rather than what is present as a malignant cause. As stated in a 1984 National Science Board report, "The nation should reaffirm its commitment to full opportunity and full achievement by all. Discrimination and the lingering effects thereof, due to race, gender, and other such irrelevant factors, must be eradicated from the American education system." [6]. Excellence and elitism are not synonymous.

Strategies for Increasing Participation

The success of educational programs in increasing students' interest and participation in science depends largely on knowing the underlying factors associated with their interest in science.

In 1985, I conducted a study to identify the factors influencing black women toward an interest and careers in science. More specifically, the objectives of the study were:

- to acquire through systematic inquiries data revealing the influencing factors that motivated minority women in general, but black women in particular, to select science as a field of study, and
- based on collective observations and insights, to design educational strategies aimed at increasing scientific knowledge, interest, confidence, and hence opportunities for many black female students.

Method and Data Source

The data collected for this study are based on respondents that participated in a 1985 survey of successful black female science professionals. This involved mailing questionnaires to 140 minority women located throughout the United States who have successfully pursued careers in science or science related fields in order to find out the factors that influenced them to select science as a field for study. Questionnaires were mailed to science professors at predominantly black colleges and universities, minority scientists employed in industry, and members of the Minority Women in Science organization as documented by AAAS. On the 36-item questionnaire, participants were asked to list and discuss those influencing factors and to rank them in order of most impact. They were also asked to make recommendations for future female minority scientists.

Results and Data Analysis

I received responses from 75 minority women professionals (72 blacks, 2 Hispanics, 1 Chinese) in 30 states and the District of Columbia. The discussion of this study is based on the responses of the 72 black women professionals. These women were employed in 28 different fields, including medicine (doctors, dentists, pharmacologists), education (college deans, professors, high school science teachers), and industry (geologists, engineers). Several of these scientists were employed in senior level positions in industry throughout the country. The level of professional degrees included bachelor's, master's, and doctorate

Table 1. Employed scientists and engineers by race and sex: 1984 [9]

Race	Total	Men	Women
Total	100%	87%	13%
White	100%	88%	12%
Black	100%	75%	25%
Asian	100%	86%	14%
Native American	100%	93%	7%
Total[1]	—	100%	100%
White	—	92%	88%
Black	—	2%	5%
Asian	—	5%	5%
Native American	—	1%	(2)

[1]Detail will not add to total because no report and other are included in the total.
[2]Less than 0.5 percent.

Table 2. Science/Engineering field distribution of women by race: 1984 [9]

Field	Total	White	Black	Asian	Native American
All scientists and engineers	100%	100%	100%	100%	100%
Scientists	86%	86%	87%	77%	87%
Physical scientists	6%	5%	5%	10%	(1)
Mathematical scientists	4%	4%	7%	2%	(1)
Computer specialists	22%	22%	25%	27%	7%
Environmental scientists	2%	2%	(1)	(1)	(1)
Life scientists	16%	16%	9%	16%	33%
Psychologists	17%	18%	19%	4%	20%
Social scientists	18%	18%	21%	18%	13%
Engineers	15%	14%	14%	23%	13%

[1]Less than 0.5 percent. NOTE: Detail may not add to totals because of rounding.

(Ed.D., Ph.D., M.D.). They graduated from college between 1933 and 1982.

The analysis of the data revealed that a variety of factors influenced these women to study science, but four factors were more influential than others. These were early exposure to science (83 percent), role models (78 percent), parental influence (78 percent), and persistence (76 percent). Other factors cited were expectations from parents, teachers, and peers (68 percent); career orientation (58 percent); and science projects (56 percent). Ten persons indicated that they selected science as a field of study because they had always had a strong interest in the subject.

The factors listed were common and systematic among all age groups. These factors correspond to those listed by Rowe as variables that might affect career selection for blacks [7]. She said, however, that without systematic research, one could only speculate to what extent these factors are relevant.

Implications and Recommendations

This study, aimed at improving black women's access to science and science related fields, has paved the way for improved understanding of the factors allowing black females to study science. All of the evidence received from the science professionals through the questionnaires and personal interviews indicates that black females will select science if the right variables are present.

Early Exposure to Science. There was a strong feeling among all of those surveyed that motivating females, particularly black females, to study science depends on our ability to interest them in science in early elementary school. Teachers of these early grades must display an active liking for an involvement in the concepts and activities of mathematics and science. From a developmental point of view, grades six to nine seem to be crucial years for developing scientific interests; the curriculum should emphasize stimulating and practical science activities. During these years, when sex difference in science and math begin to emerge, science is often taught, unfortunately, by people who might prefer to move into high school science teaching or administration.

A great deal can be done in early grades to foster basic conceptual development in science. Curriculum materials should present a larger range of occupational roles for women. The science curricula and materials should also stress the relevance and application of science content to everyday life. Career counseling to raise awareness of the need for girls to plan careers and to widen the range of career options is essential. Counselors, teachers, and parents must all re-examine the attitudes they display concerning girls' involvement in science.

Early exposure to science involves not only an introduction to science at the early elementary level, but also early exposure to museums, field trips and other community resources, reading materials in science, and role models at all professional levels. The development of study habits and test-taking skills should also begin in the early grades.

Skillful and early introduction of these variables beginning at the early elementary level is critical. This exposure should be across the spectrum, not limited to the so-called gifted and talented. Opportunity is the operative word, according to the surveyed group.

Role Models. Having early exposure to relevant role models is one of the most influential factors in sparking interest in science among black students. Therefore, it is especially important for black female students to have early and extensive exposure to black women employed in the sciences, both in teaching and in research at all professional levels. Early fostering of role model and mentor associations for black and other minority students should be developed on a planned, sustained basis through black professionals' classroom visitations, club activities, or even having them serve as project directors.

A 1979 report by the National Research Council noted that as children approach adolescence, the availability of role models becomes an important factor in their selection of future careers [6]. Rowe also cites this as a possible reason that blacks choose fields other than the sciences [7]. The presence of role models in the lives of those surveyed was the second most influential factor motivating them to select science as a field of study.

It is especially important for black female students to have early and extensive exposure to black women employed in the sciences, both in teaching and in research at all professional levels.

Parental Influence. Parents are significant figures in encouraging black students to study science. Although career and related educational preparation requirements should be communicated on a regular basis at an early age to students, it is important that their parents are involved throughout the educational program. Parents should be encouraged to actively participate in educational programs and activities for their children. Career day experiences, as well as planned academic conferences and workshops designed especially for parents should be conducted. Setting up a system of student–mentor and parent–mentor relationships with black science professionals was also

Although the group did not include guidance counselors among the major factors influencing them toward science careers, they believe that adequate counseling is an important gate to opportunity, especially for black students.

encouraged by the survey group. It is important that parents of black children see, along with their children, the relevance of science to everyday life. The more knowledgeable parents are about its relevance and the more insistent they are, the more likely their children will become interested in science.

Persistence. The black women who have been successful in science have developed the ability to persist. Having this ability in a time of conflict is essential to positive self-concept. To develop this quality, the student needs to develop an internal locus of control. Black females must learn to respond positively to intellectual challenge; they must choose to fight rather than flee. The perception among black females that science and mathematics are harder than other academic subjects and that there is little likelihood for success for them in these subjects must be eliminated. To do this, a well-structured academic development ladder that introduces science and math and practical applications early, in an educational program that builds sequentially upon the learning process, is recommended.

Counselors. Although the survey group did not include assistance from guidance counselors among their influential factors, they believe that adequate counseling is an important gate to opportunity, especially for black students. Counselors can inform students and their parents about course selection, and teachers should consult with each other prior to advising the students, in order to ensure an integrated process. Close communication between counselors, teachers, and parents through frequent group meetings is recommended. This will ensure consistency in the message relative to steps in the career ladder and integration of academic courses into a coherent academic advancement program.

Summary and Discussion

To determine the relevance of the factors identified in this study, the factors need to be introduced to black female students in an intervention or enrichment program at the early elementary level and systematically evaluated and followed throughout the educational process. Most of the programs designed to increase students' interest and participation in science are for the high school level. But programs at the early elementary level might prove to be far more effective than those at the high school level. An evaluation of existing programs designed to increase minority interest in science is needed and recommended.

Black females must be encouraged to study science and to see the relevance it has to their career plans. The number of black women pursuing science careers has increased and it must continue to do so. We need to be assured that they are well represented in the scientific community—playing a role in directing technological innovation and determining the speed and manner in which new development are introduced into society. Failure to develop and utilize the talents of certain segments of our population may have profound social, economic, and political consequences in a world where the impact of science and technology is becoming increasingly significant. □

***Julia V. Clark** is an associate professor of science education at Texas A&M University, College Station, TX 77843.*

References

1. Beane, Deanna Banks. *Mathematics and Science: Critical Filters for the Future of Minority Students.* Washington, D.C.: The American University, 1985.
2. Clark, Julia V. "The Status of Science and Mathematics in Historically Black Colleges and Universities." *Science Education* 69(5):673–79; 1985.
3. Kahle, Jane B. "What Assessment Says About Science Education for Black Students." *Phi Delta Kappan*, April 1980.
4. National Academy of Science. *Climbing the Academic Ladder: Doctoral Women Scientists in Academe.* National Research Council. Washington, D.C.: 1979.
5. National Assessment of Educational Progress. *The Third Mathematics Assessment: Results, Trends, and Issues.* Denver, Colo.: Education Commission of the State, 1983.
6. National Science Board Commission on Precollege Education in Math, Science and Technology. *Educating Americans for the 21st Century.* U.S. Government Printing Office. Washington, D.C.: 1984.
7. Rowe, Mary Budd. "The Forum: Why Don't Blacks Pick Science?" *The Science Teacher* 44:3–435; 1977.
8. *Women and Minorities in Science and Engineering.* National Science Foundation. Washington, D.C.: 1984.
9. *Women and Minorities in Science and Engineering.* National Science Foundation. Washington, D.C.:1986.

Recruiting and Retaining Talented African-American Males in College Science and Engineering

The Meyerhoff Scholars Program at UMBC—a Stunning Success

Freeman A. Hrabowski, III and Willie Pearson, Jr.

Nationally, African-Americans constitute less than three percent of the total work force in the engineering and science professions. While the nation suffers a general shortage of scientists and engineers, the low percentage of African-Americans in these professions shows that they are losing ground academically (National Science Foundation 1986).

The Meyerhoff Scholarship Program at the University of Maryland Baltimore County (UMBC) attempts to address this problem. The program is designed to increase the number of African-Americans, especially males, who enter and succeed in undergraduate and, ultimately, doctoral or professional programs in science and technology. The success of the program's first year suggests that certain initiatives, carefully orchestrated, can indeed improve that representation.

STUDIES DEMONSTRATE THE NEED FOR THE MEYERHOFF-TYPE PROGRAMS

Although more African-American males are graduating from high school than ever before, the enrollment of these students in college has been severely reduced. Their high school graduation rates increased from 20 percent to 31.9 percent between 1970 and 1985, which was the largest increase of any race-gender group (Gibbs 1988). However, the college entrance rates of African-American males declined by 6.3 percent between 1980 and 1984 (Gibbs 1988).

The students' high school experiences could explain the decline. Many African-American males get less attention and praise and more criticism than other students (McJamerson and Pearson 1989). They are seen as threatening and disruptive in the classroom and are more likely to be placed in classes for the mentally retarded or emotionally disturbed (Serwatka 1989). Many avoid advanced mathematics and science in high school, possibly because their teachers think they are unable to compete in these courses.

There is evidence that African-American males may not value education and, furthermore, are unwilling to enroll in advanced courses (Serwatka 1989).

Thus, African-American males restrict their opportunities to pursue science-related careers. Many of these students lack confidence in the power of science to solve at least some of the world's problems. They are "less convinced (than white students) of the benefits of science to society and less supportive of science research" (Anderson 1989). Three other researchers, Austin (1985), Slavin (1987) and Treisman (1985), suggest that, frequently, higher levels of achievement result when teachers, parents, and community groups encourage continued efforts and decrease isolation through support networks and role models.

This is particularly important in the education of young male African-Americans. Too often, as these students strive for academic excellence, their efforts are not reinforced, and are actually discouraged by their peer

Freeman A. Hrabowski, III is acting president, at the University of Maryland, Baltimore County, Baltimore, MD 21228. Willie Pearson, Jr., is a professor of sociology at Wake Forest University, Winston-Salem, NC 27106.

groups, both in and out of school. This, of course, makes it much more difficult to continue to try to excel in school. For whatever reasons, more white male students take college preparatory classes, especially those in science and mathematics, than do their African-American male counterparts (Carzile and Woods 1988). As a result, the number of eligible African-American male high school graduates has decreased, and many of those who do enroll in college are unprepared academically (Dyson 1989).

There are reasons other than lack of preparation that determine whether or not African-American males will attend college or whether or not they will drop out by the end of the sophomore year. For some, much depends on family income and the influence of friends with no academic goals. Others choose military careers or vocational programs because they are unaware of the benefits of a college education (Goldstein 1990). The increased use of drugs and the rising rates of incarceration are also significant (Carzile and Woods 1988). One recent study determined that more African-American males were in the criminal justice system (609,000) than were in college (436,000) (Meddis 1990).

Once enrolled, remaining in college presents other problems, not the least of which are increased tuition costs, reduced financial aid, and income lost through delayed employment (McJamerson and Pearson 1989). Two other determinants which are extremely important for retention are students' successful social integration and relationships with their instructors (Pincus and DeCamp 1988).

It is no wonder that, between 1976 and 1987, the number of bachelor's degrees earned by African-American males fell 12.2 percent, and the number of master's degrees earned fell 34 percent (Carter and Wilson 1989). Indeed, many college graduates decide not to pursue graduate degrees. Of the 33,456 doctorates conferred in 1988, only 2.8 percent were awarded to African-American males (McJamerson and Pearson 1989). In fact, fewer than three percent of doctorates in science and engineering are awarded to African-Americans annually. In that year, they earned 46.7 percent fewer Ph.D.s than they did in 1978, which was the largest decline of any race-gender group (Carter and Wilson 1989). And furthermore, in 1986, over half the doctorates earned by African-American males were in education-related fields. Engineering and the physical sciences ranked lowest.

Michael Hirschorn attributes this low level of achievement to the shortage of well-prepared high school students, the decreasing number of African-Americans enrolled and graduating from college, and the lack of guidance in college (Hirschorn 1988). He also cites the heavy competition from American industries for minority college graduates, especially those with engineering degrees. College and university science faculties grapple with ways to increase the numbers of minority students in the sciences, but few have documented statistics on those majors who succeed and actually graduate each year. Most faculties only sense that the numbers are slight for those who earn A's or B's in freshman calculus, chemistry, physics, biology or engineering.

UNIVERSITY OF MD—BALTIMORE COUNTY

Meyerhoff Program students at the University of Maryland, Baltimore County.

■■■

UMBC has studied the problem, and has discovered that, in 1990, for example, of the more than 1,300 bachelor's degrees awarded, 23 percent were earned by white students in science and technology and fewer than two percent earned by African-Americans. Furthermore, many of the latter science majors' grade point averages were below B, and thus, they were ineligible for graduate school.

The Meyerhoff Scholarship Program

We need to insure that young African-American males have a place in college and in the world of science and technology. The severe shortage of African-American scientists suggests that even the most talented of students are not succeeding in these fields. To increase these numbers, we must identify high achieving students and provide them special academic and emotional support.

With these goals in mind, UMBC initiated the million-dollar Meyerhoff Scholarship Program. By 1987, Maryland's overall African-American college enrollments reflected the national trend, and were in decline, having dropped from 43 percent to 37 percent in a decade. The following year, UMBC approached the Robert and Jane Meyerhoff Foundation to explore its interest in reversing this trend and, especially, to help stem the declining numbers of African-American males entering the state's colleges and universities.

The Foundation provided more than $500,000, matched in part by the university, to establish what has now become the multi-million dollar Meyerhoff Scholars Program. In 1989, 10 outstanding Maryland high school students were named Meyerhoff Scholars. These students received full four-year scholarships, a personal computer, and special academic support, including a precollege summer program, academic advising, special counseling, and tutoring. They attended special lectures, took science and cultural field trips, and, the following summer, were to receive summer internships that focus on science and technology.

Nine additional students, designated as finalists, also participated in the program and received major scholarships that are provided through University funds.

The African-American students' high school experiences could explain the decline in college enrollment. Many African-American males get less attention and praise and more criticism than other students. They are seen as threatening and disruptive in the classroom and are more likely to be placed in classes for the mentally retarded or emotionally disturbed.

STUDENT PROFILE

In assessing the effectiveness of the Meyerhoff Program, it is important to look at the academic and personal backgrounds of the participants. The 19 UMBC freshmen were selected because their high school grades, standardized test scores, letters of recommendation, essays, and personal interviews pointed to their interests in and strong aptitude for work in science and technology. The Scholars' mean SAT score was 1196 and they earned a high school GPA of 3.8 on a 4.0 scale. The Finalists' mean SAT score was 1118, and their GPA was 3.5. All of the students were in the top 5 percent of their class and several were valedictorians, salutatorians, National Achievement finalists, Maryland Distinguished Scholars, and honor society students.

They had been accepted at Stanford, MIT, Yale, Cornell, Johns Hopkins, and other nationally eminent universities. They selected the Meyerhoff Program because they wanted to be associated with other talented high-achieving African-American males, and because of the level of research and the financial and academic support the program offered. Approximately 90 percent of the students lived in two-parent families. The others lived with one parent, usually their mothers. About three-fourths of the scholars had always lived with their biological fathers.

In general, Meyerhoff Scholars came from small nuclear families. Two-thirds had only one sibling, usually a sister. Slightly more than half were only children or first-borns. More than likely it was their mothers, not their fathers, who had read to them when they were young. Half of the parents were college graduates. Many had master's degrees and some had doctorates. Overall in the group, more fathers held professional positions than mothers. Not surprisingly, most students had home computers, encyclopedias, VCRs, and regular subscriptions to magazines and newspapers. What is surprising is that two-fifths of the students reported that their parent or parents never attended PTA meetings or parent-teacher conferences.

Students ranked religion or spirituality "important" or "very important" to their success, and a majority attended church regularly while in high school. They considered as important correcting social and economic inequities; community leadership; a steady income; living close to parents and relatives; and having children and being able to provide better opportunities for them. They considered as very important finding the right person to marry. Most thought that having a lot of money was somewhat important. Almost everyone placed a high value on strong friendships and leisure time for self-enjoyment. Half the students attended predominantly white high schools; 30 percent attended predominantly black schools; and the remainder attended racially-balanced schools. While in school, they spent two or more hours daily on homework and read newspapers, books, and magazines on a regular basis. Most of the students received some career and academic counseling, usually from their parents, teachers and counselors. Few held part-time jobs, and most used their allowances to purchase reading materials.

These students participated in community activities, athletics, debate, drama, student government, honor societies, math/science clubs, and black student organizations. Fifteen percent of them were members of bands. About 80 percent of the students considered themselves better in math than most of their peers, and half thought of themselves as better in science. Nearly everyone wanted to earn a doctoral degree or its equivalent.

Summer Bridge Program

These, then, were the 19 students who entered the six week residential summer bridge program at UMBC in 1989. They took credit courses in mathematics and African-American literature and noncredit courses in chemistry and study skills.

The math and chemistry courses and regular special sessions strengthened their grasp of concepts they had been exposed to in high school and introduced several new concepts emphasizing problem solving. The literature course reinforced their reading and writing skills and gave them a broader understanding of African-American culture. The study-skills course focused on note taking and time management and other techniques needed to succeed in college.

Parents were invited to attend a variety of activities and show their support. During the summer, students went on picnics, played sports, and took trips to museums, plays, national scientific agencies, and engineering firms. Also, during that time, each student was assigned an African-American male whose profession was science or technology.

Faculty, staff, and student evaluations of the first summer bridge program have lead to several changes. During the first summer, for example, all students were placed in the same mathematics course. We found that the quality of precalculus and calculus courses varies widely among high schools, so, in the future, students will be placed according to their score on a university calculus readiness test. We cut the number of summer field trips and extra-curricular activities to give the additional study time the students said they wanted. In 1989, there were 15 activities; there were nine in 1990.

The Scholars felt that several program components were very helpful. Mathematics and chemistry problem-solving sessions enabled students to work together, to ask more questions, and to attempt more difficult problems. They found also that studying as a group was extremely helpful. Toward the end of the summer, program staff worked with the students to select courses for the fall. Most were advised to take no more than two courses in mathematics, science, or engineering, and no more than 18 credit hours. As in the summer, students were encouraged to study in small groups and work actively with tutors. Every week during the summer and every other week during the first semester, students met with the director or the coordinator of the program to discuss general concerns related to the classroom and to extra-curricular activities.

First Academic Year

Results of the first semester were very positive. The Scholars' mean GPA was 3.7; the Finalists' was 3.1. Nine Scholars and six Finalists made an A or B in all mathematics and science courses. For the year, the students' overall GPA was approximately 3.5; all students earned at least 3.0. By comparison, only 12 percent of all freshman at the University earned a 3.5 GPA or better. During the first year, students completed an average of 32 credits, including seven credits for the summer program.

Examples of Mathematics and Science Coursework Taken by Meyerhoff Scholars and Finalists During the First and Second semesters, and average GPA's

Course	GPA
Chemistry I (15 students)	3.33
Chemistry II (10 students)	3.3
Engineering (10 students)	3.6
Calculus I (19 students)	3.53
Calculus II (9 students)	3.44
Biology (5 students)	3.2

In each of these courses, more than 25 percent of all UMBC students failed.

All of the Scholars were enrolled in the UMBC Honors College, which provides an enriched education for gifted students. UMBC Honors Students have either a secondary school GPA of 3.5 or higher, and a minimum combined SAT score of 1000 or a combined SAT score of 1200. During the first year of the program, Meyerhoff students enrolled in the honors sections of chemistry, biology, calculus, and French. Eleven of the students during the first semester and twelve in the second semester achieved academic honors, requiring a GPA of 3.5 or higher, and enrollment in at least 12 hours.

In each semester, four students earned a C in a course. Special individual and group sessions were held with these students, each of whom said he/she had not worked up to his/her potential and needed to study harder. In most instances, students had not worked with other students, black or white, and had not realized the need to do so until it was too late. It is noteworthy that two of these students had inconsistent high school grades, receiving A's as well as C's while receiving in excess of 700 on the Math SAT.

Program staff will continue to recommend that students take no more than two mathematics and science courses during the first semester. We will encourage students who earn a C in a course to repeat it in order to develop a strong foundation before advancing to the next higher-level course. Repeating a course also encourages the development of greater self-esteem, as demonstrated by the four students who did repeat courses. They improved their grades to A or B.

The Meyerhoff Scholars not only earned solid academic records but also adjusted well to college life. Participation in the Summer Bridge Program allowed them to experience college-level mathematics and science coursework and testing before they entered their freshman year. Moreover, they developed a strong peer-support group which carried into the academic year when individual and group discussions of academic performance and college life became even more important. They were also encouraged to work together in small groups and to get to know tutors well before they were needed.

The students' academic performance demonstrates that these strategies worked well. It appears that their high school academic performance, particularly in math and science, was an important predictor of college success. All the Scholars and Finalists returned for their second year, and a second group of Scholars, seven men and seven women, began their college work in the summer of 1990. Two additional out-of-state students, one male and one female, began this fall.

What We Can Conclude from This Study

One might expect that these Meyerhoff students would have succeeded even if the program had not provided special academic support. However, a review of the records of all African-American science and engineering majors at the UMBC over the past five years found that even students with similar backgrounds had earned one or more D's in freshman math and science courses. The shortage of scientists and engineers is in part explained by the movement of students, white and African-American, from majors in science and engineering to other fields because of poor performance in freshman level math and sciences courses.

It is clear that initiatives like the Meyerhoff Scholarship Program, which first focuses special attention on performance in a prefreshman summer program and on special tutoring, decision making, and group study, can lead to increased numbers of minority students who successfully pursue careers in science and engineering. ❑

Appendix
Survey Results—Personal Characteristics of Meyerhoff Participants:

Characteristic	Percent
two parent families	90%
always lived with biological father	75%
one sibling	66 2/3%
first born child	33%
parents holding bachelor's degrees	50%
parents holding master's degrees	40%
parents holding doctoral degrees	10%
fathers holding professional positions	65%
mothers holding professional positions	45%
students having PCs, VCRs, encyclopedias, etc.	70%
religion as an important factor in lives	80%
regular church attendance	63%
marriage to "right person" thought very important	75%
graduated from predominantly white high school	50%
career and academic counseling:	
from mother or teacher	80%
from father or counselor	70%
considered math ability better than that of peers	80%

References

Anderson, Bernice. 1989. Black participation and performance in high school science. *Blacks, Science, and American Education.* Willie Pearson, Jr., and H. Kenneth Bechtel, editors. Rutgers University Press, p. 47.

Austin, Gilbert. 1985. *Research on Exemplary Schools.* (H. Garber, co-editor). New York: The Academic Press.

Carter, Deborah and Reginald Wilson. 1989. *Minorities in Higher Education.* American Council on Education, Office of Minority Concerns.

Carzile, Samuel and Jacqueline Woods. 1988. Strengthening black students' academic preparedness for higher education. Journal of Black Studies 19: 150–162.

Dyson, Michael. The plight of black men. *Zeta Magazine* Feb. 1989: 5.

Gibbs, Jewelle, ed. 1988. *Young, Black and Male in America: An Endangered Species.* Auburn, pp. 76, 87–88.

Goldstein, Amy. Effort urged to put black males in college. *The Washington Post* 14 Jan. 1990, A9.

Hirschorn, Michael. 1988. Doctorates earned by blacks decline 26.5 in decade: 820 degrees awarded to them in 1986, over 50 to Women. *The Chronicle of Higher Education.*

McJamerson, Evangeline and Willie Pearson, Jr. 1989. The declining participation of African-American males in higher education: Causes and consequences. Paper prepared for Mid-South Sociological Association in Baton Rouge, Louisiana.

Meddis, Samuel. Young black 'generation' in legal webb. *USA Today* 27 Feb. 1990, 3A.

National Science Foundation. 1986. *Women and Minorities in Science and Engineering.* Washington, D.C.

Pincus, Fred and Suzanne DeCamp. Sept. 1988. Minority Community College Students Who Transfer to Four-Year Colleges: A Study of a *Matched Sample of B.A. Recipients and Non-Recipients.* p. 5.

Serwatka, Thomas. 1989. Correlates of the underrepresentation of black students in classes for gifted students. *The Journal of Negro Education.* Fall issue.

Slavin, Robert E. 1987. Grouping for instruction: Equities and effectiveness. Office of Educational Research and Improvement. U.S. Department of Education. OERI-G-86-0006.

Treisman, Philip Uri. 1985. "A Study of the Mathematics Performance of Black Students at the University of California, Berkeley." University of California at Berkeley.

Asian-American Students

The Myth of a Model Minority

The extraordinary success some Asian Americans are enjoying in a few areas has created the misconception of an exemplary minority and has detracted from a need to accurately assess the costs and sacrifices involved for those who manage to reach such high levels of achievement.

Frank H. Shih

The educational achievements of Asian Americans have recently generated much publicity. News articles have repeatedly directed our attention to the outstanding academic successes of this group. For example, we read of the disproportionately large number of Asian Americans reaching the finals of the prestigious Westinghouse Science Talent Search and of others who, despite having to catch up in English, become high school or college valedictorians [3,4,9]. It has also been reported that Asians as a group have the highest score on the math section of the Scholastic Aptitude Test while taking it at a rate greater than any other group. The capstone on these accomplishments, however, is the growing rate of Asian enrollment at top universities: 14 percent at Harvard, 20 percent at MIT, and 25 percent at Berkeley, percentages so dramatic that many are afraid schools have placed limits on Asian admissions [2,4].

These "new whiz kids," as *Time* magazine recently called them [2], have been touted as a modern success story in the old American tradition. We are told that their penchant for hard work and their drive for excellence have enabled them to overcome great obstacles in a country many have only recently called their own. The successes in education appear to have paid off. According to the U.S. Census Bureau, Asian/Pacific Islander households have the highest average and median incomes of all ethnic categories, including whites [16].

These achievements have led to the perception of Asian Americans as the model minority. This generalization, implying that all are high achievers, has not been dampened by indications that recent immigrants are from better-educated classes and thus tend to hold high educational aspirations for their children. Neither has it been emphasized that the notion of success is narrowly defined, focusing almost exclusively on the group's performance in specific areas, such as science and technical fields, and on test scores and percentage of enrollment in certain schools. Household income may also be a poor indicator if one accounts for the number of wage earners in each family or the total hours worked—both probably high for Asians.

There is also an assumption that life is going extremely well for Asian Americans on college campuses, and a new racial stereotype has emerged of the dedicated, disciplined, and very bright Asian student who has neither extracurricular interests nor social needs. Moreover, they are perceived to be content with academic accomplishments and with college life.

These images, as positive as they may seem, impact negatively on this group. It is unfortunate that the extraordinary successes some are enjoying in a few areas have brushed aside any discussion of their unique problems and have detracted from a need to accurately assess the costs and sacrifices involved for those who manage to reach such high levels of achievement.

At the State University of New York at Stony Brook, for instance, Asians in this year's freshman class received the highest math scores on the SAT's but had the lowest in the verbal section compared to other ethnic groups. A study at UCLA of Chinese students has documented that those with low verbal scores compensate for the deficiency in English by carefully avoiding classes that appear to require a high degree of language skills [14]. Besides limiting access to the full range of courses that the university offers, researchers also found that the failure to develop English skills forces these students to spend longer hours on their homework [14].

There are important social ramifications as well. This deficiency contributes, for instance, to the tendency of Asian Americans to isolate themselves by socializing only with those who speak their own languages. It also increases the reluctance to take advantage of student services and participate in social activities. And, finally, the lack of adequate communication skills will be a tremendous handicap when they enter the job market.

It is clear that not all Asian-American college students are underprepared in English. But for those who are, limited proficiency in English exacerbates the alienation and cultural margination that many Asian Americans feel to a degree. The college campus, after all, is a cultural environment that is vastly different from that of an average Asian-American home. Adjusting to campus life—from norms regarding male/female relationships to the food served in the cafeteria—is thus more difficult for them than for the traditional white student. The dissimilarity between the two cultures, however, is more subtle than can be deduced from behavioral patterns.

The principles governing social conduct in Asian societies is fundamentally different from those of Americans. For Asians, social interaction must always be congenial, and behavior is carefully crafted towards maintaining group harmony. This so-called "situation centeredness" [5] contrasts with American emphasis on individualism. For example, while Asian mannerisms and speech express self-deprecation and deference to others, these traits are seen as weaknesses in a culture that emphasizes looking out for number one and the importance of asserting one's opinions. The differences are such that even with fluency in English, social interaction for many Asian Americans may involve a certain amount of anxiety.

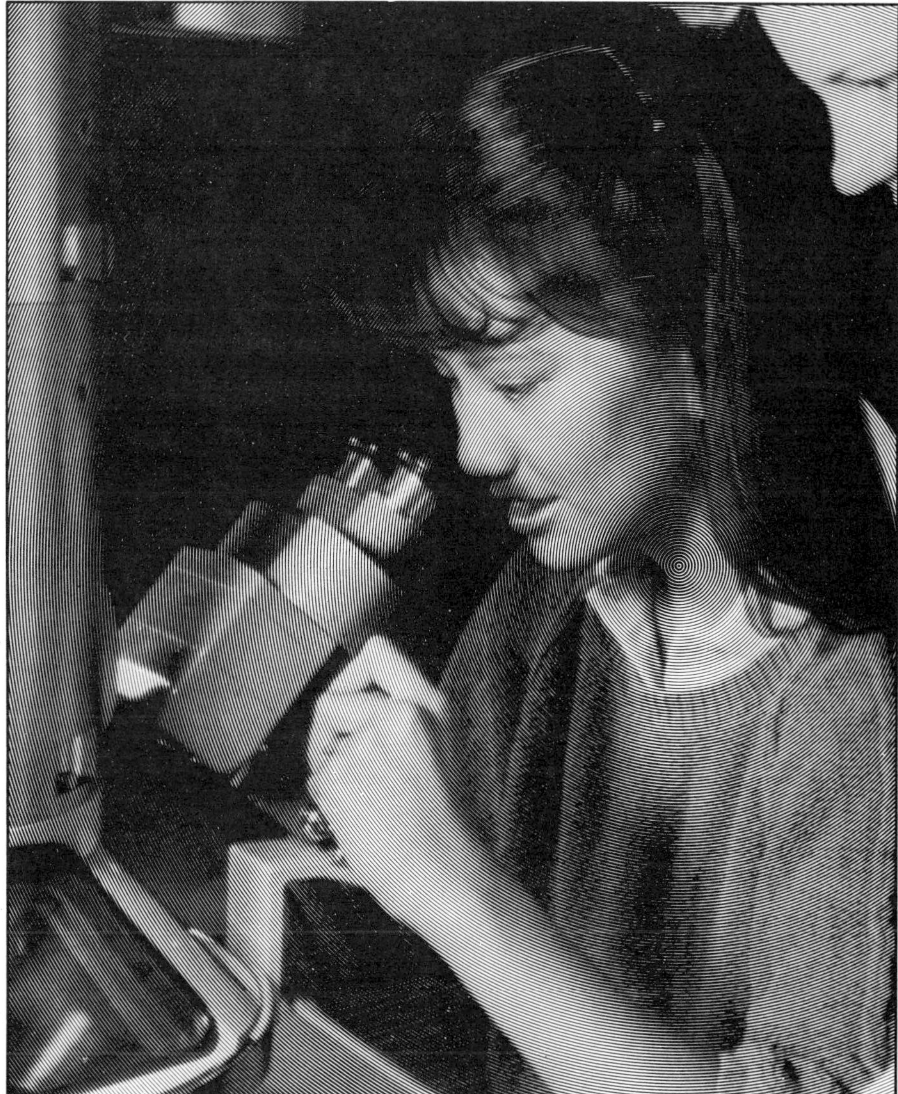
UNIVERSITY OF MARYLAND

Family pressures add another dimension. This is felt in two ways. In Asian societies, a person's social identity is closely associated with the kin group, an identification that is fostered by a strong sense of shame and guilt [15]. Asian children are taught that they bear the responsibility for the honor or shame of each member of their family. Because success or failure is shared with one's family, the consequences of one's actions are greatly magnified, making success more imperative and the fear of failure more intense. A second form of family pressure can be seen as a corollary of the first. Though the tight-knit family structure deserves much credit for the group's achievements, it should also be mentioned that parents' expectations are high. Moreover, career goals are usually determined by parents. Thus, realizing these expectations—sometimes an awesome task in itself—is made more difficult by having to do so in a field that is not of the student's choosing.

How do Asian-American students under these circumstances cope with the emotional, social, and academic stresses? Are there psychological consequences? The evidence we have indicates that, like the larger population of Asian Americans, these students are experiencing greater levels of emotional and social adjustment difficulties and have a higher rate of unreported mental health problems than do Caucasians [11,12,13,17]. Not surprisingly, assimilation has some effect: those who are recent immigrants tend to have a greater degree of unhappiness and anxiety [7,12]. However, research has shown that Asian values are retained even among those relatively assimilated and that all Asian students may at one time or another experience similar forms of distress [8].

Racial visibility is another factor that compounds the problem. Shedding the foreigner image and gaining acceptance by the majority is difficult for Asian Americans because of it. The extent to which this creates cultural identity problems for those who have been Americanized remains to be fully explored [10].

Because in Asian cultures the body is a legitimate vehicle for articulating psychological problems, emotional suffering is often concealed. Asian students who complain of physical ailments should be asked directly if there are emotional problems.

Some Suggestions

There are a number of ways professors and other college professionals can respond to the needs of Asian-American students. Because a large number of these students major in the sciences and mathematics, it would seem appropriate for educators in these fields to take the lead and use their influence in the role of advisors on personal as well as academic issues. The following suggestions can be helpful:

- Question the student's choice of course electives. For example, those with limited conversational English could be reminded to take English as a Second Language courses. In classes that are attended in large numbers by Asian Americans who are non-native speakers of English, professors could emphasize that proficiency in writing and reading are necessary not only in course work but also especially when one begins to look for employment and later pursues promotions.
- Encourage participation in social activities, as a way of improving one's conversational English and meeting new friends.
- Refer students to the campus counseling or psychological center when signs of emotional strain are apparent. In many instances, merely suggesting that the person seek help will not be enough, as students feel intimidated by the prospect of having to articulate personal problems to someone outside the family. The feeling of being stigmatized and, in particular, of bringing disgrace to one's family will also be a problem. In addition, for therapy to be successful, the counselor will have to be sensitive to cultural issues as well as Asian perceptions of the therapeutic experience [14].
- Address the issue of academic pace. When a course load appears to be too demanding, ask the student to consider reducing the amount of credits. The value of education in Asian societies is such that these students feel compelled to maximize semester credits and not "waste" the opportunity.
- Inquire about the student's physical well-being and take an interest in his or her study/exercise/eating routine. Under extreme pressure to do well, students may sacrifice good health habits. In addition, Asian Americans, especially those who are recent immigrants, have a hard time adapting to non-Asian meals.

Professors should also inform those who provide campus services of the particular needs of Asian Americans and the way such needs are met. The campus infirmary, for example, must understand that traditional Asian attitudes concerning medicine influence the way psychological problems are expressed. Because in Asian cultures the body is a legitimate vehicle for articulating psychological problems, emotional suffering is often concealed [7]. Asian students who complain of physical ailments that are not supported by a physical examination should be asked directly if there are emotional problems.

Research also shows that Asians perceived counselors in their own racial group to be more credible [1]. There should be a concerted effort by administration and faculty to hire Asian student and professional staff proportional to the student population.

Conclusion

Asian-American students are a heterogeneous group. The diverse cultural and linguistic representation and differing degrees of adaptation to American life complicate efforts to assess their needs. Some of the problems many of them face are unique, while others, though common to all college students, are made more acute by their distinct cultural heritage. Those who are recent immigrants, especially on campuses where Asian students are relatively few, are most vulnerable to socio-emotional distress. It is with this group that college and university administrators should be most concerned.

Racial stereotypes, even those meant to be complimentary, tend to dehumanize individuals.

Racial stereotypes, even those that are meant to be complementary, tend to dehumanize individuals. In the effort to meet the needs of Asian Americans, faculty as well as administrators should be aware that the current misconception of Asians as a model minority obscures some of the needs of this group while allowing other problems to be overlooked. This image of success may in fact be altogether false when all the facts are known [6]. Nevertheless, we do know that high grade point averages do not translate into high salaries—not to mention successful careers, when one is unable to communicate effectively. Academic achievements, even in more than one area, will not guarantee happiness when the individual is uncomfortable with the cultural environment.

For Asian Americans, a college campus remains an excellent place to learn the skills to prepare for a rewarding career. For recent immigrants, it is ideal for acquiring language skills and for broadening their understanding of their adopted culture. A college education then, should prepare the students to participate fully in all aspects of American life. This will assure all of the benefits that come from the rich blend of two distinct cultures—Asian and American. □

Acknowledgements

The author would like to thank John McCann, Alice Shih, and James Nobles for their comments and suggestions.

References

1. Atkinson, Donald R., M. Maruyama, and S. Matsui. "Effects of Counselor Race and Counseling Approach on Asian-Americans Perceptions of Counselor Credibility and Utility." *Journal of Counseling Psychology* 25(1):78–83; 1978.
2. Brand, David. "The New Whiz Kids." *Time* 130(9):42-51; August 31, 1987.
3. Browne, Malcolm, "A Look at the Success of Young Asians." *New York Times* March 25, 1986: C3.
4. Butterfield, Fox. "Why Asians Are Going to the Head of the Class." *New York Times—Education Supplement* August 3, 1986: 18–23.
5. Hsu, Francis. *Americans and Chinese: Passage to Differences*, 3rd ed. Honolulu: The University Press of Hawaii, 1981.
6. Kim, Kwang Chung, and Won Moo Hurh. "Asian Americans and the 'Success' Image: A Critique." *Pacific/Asian-American Mental Health Research Center Research Review* 5(1/2):6–9; 1986.
7. Kleinman, Arthur. *Patients and Healers in the Context of Culture: An Exploration of the Borderlands Between Anthropology, Medicine, and Psychiatry.* Berkeley, Cal.: University of California Press, 1980.
8. Minatoya, Lydia Y., and William Sedlacek. "Another Look at the Melting Pot: Perception of Asian-American Undergraduates." *Journal of College Student Personnel* 1:328–36; July 1981.
9. Quindlen, Anna. "The Drive to Excel." *New York Times Magazine* February 22, 1987: 32–39.
10. Saul, Tuck. "Are Asian-American College Students Trying to Pass for Whites?—Conditioning in the Heartlands of America." *Asian-American Psychological Association Journal* 8:19–21; 1983.
11. Sue, Derald, and Barbara A. Kirk. "Asian-Americans: Use of Counseling and Psychiatric Services on a College Campus." *Journal of Counseling Psychology* 22(1):84–86; 1975.
12. Sue Stanley, and Derald Sue. "MMPI Comparisons Between Asian-Americans and Non-Asian Students Utilizing a Student Health Psychiatric Clinic." *Journal of Counseling Psychology* 21(5):423–27; 1974.
13. Sue, Stanley, and J.K. Morishima. *The Mental Health of Asian-Americans.* San Francisco: Jossey-Bass, 1982.
14. Sue, Stanley and Nolan Zane. "Academic Achievement and Socioemotional Adjustment Among Chinese University Students." *Journal of Counseling Psychology* 32(4):570–579; 1985.
15. Toupin, Elizabeth Sook Wha Ahn. "Counseling Asians." *Pacific/Asian-American Mental Health Research Center Research Review* 3(1):9–11; 1984.
16. U.S. Bureau of the Census. *1980 Census of Population—Characteristics of the Population.* U.S. Department of Commerce, Washington, D.C., PC 80-1-D1-A:475–86,501–12; 1981.
17. Wen, H. Kuo. "Prevalence of Depression Among Asian Americans." *Journal of Nervous and Mental Disease* 172(8):449–57; 1984.

Frank Shih is the Assistant Director, Advancement on Individual Merit Program, State University of New York at Stony Brook, Stony Brook, NY 11794-3375.

Underrepresentation of Hispanic Americans in Science

A smaller percentage of Hispanics enter higher education than other ethnic groups. In order to reverse this trend, steps need to be taken early in the education of students. Students at risk need support from a variety of sources, including the home environment.

Steven J. Rakow and Andrea Bermudez

During the past decade there have been significant efforts to increase the participation and achievement of minority students in scientific careers. Yet Hispanic Americans continue to be underrepresented in these fields [2,4,17,18], achieve at a lower level than their Anglo peers on general academic tasks [11,16], and are less likely to complete their formal educations [11,13,20].

Participation in Scientific Fields

Hispanic-American students are underrepresented both in scientific careers and post-secondary preparation for scientific careers. A 1984 NSF report indicates that Hispanic Americans represent slightly over 2 percent of all scientists and engineers and that they have less experience, lower pay, and are more likely to be working in non-science and non-engineering positions than are other ethnic groups [21].

Data on minority enrollments in higher education for 1984 show that Hispanic Americans make up 4.6 percent of the undergraduate enrollment and 2.9 percent of the professional enrollment [19]. More specifically, of the Mexican Americans and Puerto Ricans who enter graduate or professional school, 52 percent drop out before completing their degrees [5]. Of the bachelor's degrees awarded during the 1980–81 school year in the biological sciences, 2.6 percent of the degrees went to Hispanic Americans, and, in the physical sciences, 1.7 percent of the degrees were awarded to Hispanic-American students.

Several factors are suggested that might contribute to these low enrollment figures. First, a smaller proportion of Hispanic Americans enters higher education than other ethnic groups. Using the number of students entering the first grade as a baseline, 42 percent of Anglo students, 25 percent of American Indians, and 36 percent of black students enter higher learning, but only 23 percent of Mexican American and 16 percent of Puerto Rican students continue their educations past high school [9]. In 1979, among persons 25 or older, only 4.3 percent of Mexican Americans, 4.2 percent of Puerto Ricans, and 13.9 percent Cubans had completed college [7]. Hence, these students cannot be directed into careers in science and technology if they do not remain in school or if they are already behind before entering school [3].

Hispanic students have a higher rate of attrition (31 percent by age 18) than blacks or whites (20 percent and 13 percent respectively) [1]. Family background factors and early academic failure appear to play an important role in students' decisions to leave school [20]. Minority students also express lower aspirations and expectations than white students, with Hispanic students having lower expectations than black students. Of Hispanic-American groups, Cuban students have the best self-image and most confidence, while Puerto Rican students have the lowest aspirations and expectations [8,11].

Thus, there appears to be a clear pattern of underrepresentation in science and technology for Hispanic-American students. The underrepresentation has been associated with several classroom-related variables. Most important appears to be the rate of attrition; if students are not in school, they cannot be expected to succeed in science and technology careers.

Olstad, *et al.* summarized a number of factors that might contribute to the underachievement of ethnic minorities in general in the fields of science and mathematics [12]. These factors include testing, learning characteristics, classroom experiences, and counseling experiences. Malcom has suggested additional factors that serve as barriers to the participation of minority students in science, including lack of skill in English, tracking in non-academic programs, lack of continuity due to migration, cultural values not supporting a choice of science, peer pressures not to study, inappropriate use of tests that result in labeling non-English language students as retarded, inadequate preparation in science and mathematics, and lack of role models [9].

Hispanics are just as motivated to pursue further academic study. However, they complete fewer science courses in junior and senior high school. In general, Hispanics tend to come from lower income homes and have fewer science-related experiences.

Achievement of Hispanic Americans in Science

The most comprehensive examination of the achievement of Hispanic-American students in science is based on a series of national assessments. Stratified random samples of 9-, 13-, and 17-year olds were tested in 1969–70, 1972–73, 1976–77, and 1981–82.

Olstad, *et al.* have summarized the results of the 1976–77 National Assessment of Educational Progress in Science for white, black, and Hispanic students [12]. On cognitive items, white students consistently score above the national mean, while black and Hispanic students scored below it. Hispanic students, however, consistently out-performed black students on these items. Results of the 1981–82 national assessment reveal the same order of achievement: white, Hispanic, black [14,15].

Robinson, Gerace, and Mestre compared factors influencing the science achievement of Hispanic and Anglo university technology students [16]. They examined the relationships among academic preparation, motivational factors, socio-economic status, and science achievement. Students in their study showed no difference in their motivation for pursuing further academic study, although the groups did differ in the number of science courses they had completed in junior and senior high school. The largest difference, however, was found on socio-economic variables. Hispanic students generally came from homes with a lower average income and also had significantly fewer science-related experiences.

Conclusion

The most striking observation from this review is how little is known about the status of Hispanic-American students in science. While these students comprise America's youngest and fastest growing population [10], little is known about their achievement in and attitudes toward science. More important is the lack of information regarding the factors that contribute to the success or failure of Hispanic Americans in science. Yet the failure to attract and retain Hispanics in scientific fields is well documented. Home factors appear to play an important role, but the influence of ethnic culture is only beginning to be explored.

Our society is becoming increasingly dependent upon scientific and technological skills. Whereas only a few decades ago, literacy was defined as the ability to read and write, more and more it is coming to mean the ability to use the tools of science and technology in our everyday lives. In the same way that those denied the chance to learn to read and write are severely handicapped, so too, those denied access to scientific and technological literacy also will be disadvantaged.

Steps to reverse these trends must be taken early in the educations of Hispanic students. Students who are at risk need support from a variety of sources including the home environment. Beyond the technological alienation and long-term joblessness these individuals experience, there are serious labor market consequences. For example, in 1981, $228 billion were forgone in national income and $68 billion in government revenues due to students dropping out of school [6].

Rodriguez and Gallegos have suggested several steps that might be taken at the precollege level to increase the participation of Hispanic-American students in science:

- public school science curriculums organized around the science of everyday experience;
- bilingual science and mathematics instruction;
- concentrated attention to mathematical deficiencies;
- incorporation of ethnoscience;
- taking advantage of work experience; and
- academic and career counseling.

In addition to these practical measures, it is obvious that more research needs to be done not only on Hispanic-American students in the science classroom but also in an attempt to understand the kinds of cultural factors that influence the science-related attitudes and achievements of Hispanic Americans. It is only when science education seeks to meet the specific needs of each student that it can aspire to be exemplary for all students. □

Steven J. Rakow (top) is an assistant professor and Andrea Bermudez (bottom) is a professor in the School of Education, University of Houston-Clear Lake, Houston, TX 77058.

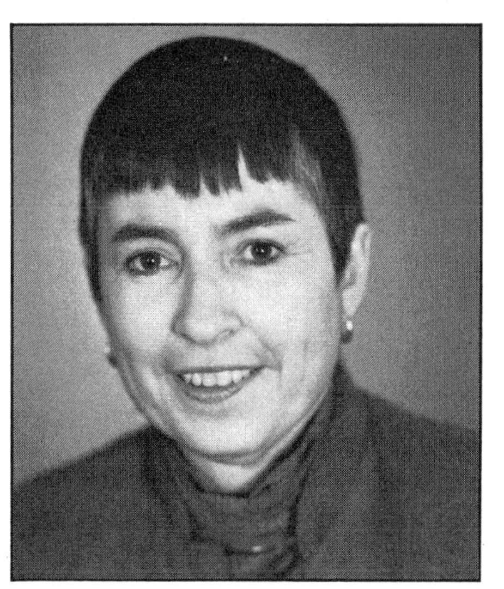

References

1. Astin, A. *Minorities in American Higher Education.* New York: Jossey-Bass, Inc., 1982.
2. Barbarosa, P. "Underrepresentation of Minorities in Biological Sciences." *Bioscience* 25: 319–20; 1975.
3. Boyer, E. *High School: A Report on Secondary Education.* New York: The Carnegie Foundation, 1983.
4. Burns, M., W. Gerace, J. Mestre, and H. Robinson. "The Current Status of Hispanic Technical Professionals: How Can We Improve Recruitment and Retention." *Integrated Education* 29:49–55; 1982.
5. Commission of the Higher Education of Minorities. *Final Report.* Los Angeles: Higher Education Research Institute, Inc., 1982.
6. Cotterall, J. *On the Social Costs of Dropping-Out of High School.* Stanford, Cal.: Stanford Education Policy Institute, Stanford University, 1985.
7. Duran, R. *Hispanics' Education and Background: Predictors of College Achievement.* New York: College Entrance Examination Board, 1983.
8. Fleming, L. *Parental Influence on the Education and Career Decisions of Hispanic Youth.* Washington, D.C.: National Council of La Raza, 1982.
9. Malcom, S. In *Puerto Ricans in Science and Biomedicine. Report of a Conference.* Washington, D.C.: American Association for the Advancement of Science, 1981.
10. National Commission on Secondary Education for Hispanics. *Make Something Happen.* Washington, D.C.: Hispanic Policy Development Project, 1984.
11. Nielsen, F., and R. Fernandez. *Achievement of Hispanic Students in American High Schools: Background Characteristics and Achievement.* Chicago: National Opinion Center, 1981.
12. Olstad, R., J. Juarez, L. Davenport, and D. Haury. *Inhibitors to Achievement in Science and Mathematics by Ethnic Minorities.* Seattle: University of Washington, College of Education, 1981. ERIC Document Reproduction Service No. BE 10278.
13. Orum, L. *The Educational Status of Hispanic American Children.* Washington, D.C.: National Council of La Raza, 1982.
14. Rakow, S. "The Status of Minority Students in Science." *Urban Education* 20:103–13; 1985.
15. Rakow, S., and C. Walker. "The Status of Hispanic American Students in Science: Achievement and Exposure." *Science Education* 69:557–65; 1985.
16. Robinson, H., W. Gerace, and J. Mestre. "Factors Influencing the Performance of Bilingual Hispanic Students in Math and Science Related Areas." *Integrated Education* 18:38–42; 1980.
17. Rodriguez, R., and R. Gallegos. *Hispanics, Engineering, and the Sciences: A Counseling Guide.* Las Cruces, N.M.: Educational Resources Information Center (ERIC), Clearinghouse on Rural Education and Small School (CRESS), 1981.
18. Sie, M., B. Markham, and S. Hillman. "Minority Groups and Science Careers." *Integrated Education* 16:43–46; 1978.
19. Snyder, T. *Digest of Educational Statistics.* Washington, D.C.: U.S. Department of Education, Office of Research and Improvement, 1987.
20. Steinberg, L., P. Blinde, and K. Chan. *Dropping Out Among Language Minority Youth: A Review of the Literature.* Los Alamitos, Cal.: National Center for Bilingual Education, 1982.
21. *Women and Minorities in Science and Engineering: Executive Summary.* Washington, D.C.: National Science Foundation, 1984.

The Need for Strengthening Native American Science and Mathematics Education

As the most underrepresented minority in the sciences, native Americans need many more role models and culturally sensitive teachers and curricula.

Gary G. Allen and Owen Seumptewa

Native Americans are the most underrepresented minority in scientific and technical professions [8]. This condition exists despite the overwhelming need in almost all Indian communities for individuals trained in the sciences to help address the problems of poverty, poor health, unemployment, and inadequate resource management. To understand why so few Indian students select science careers, you only need to begin by addressing the well-documented difficulties they experience in traditional science classrooms, especially their low achievement in comparison to other minorities and white students [1,2,3,5,8,9,21]. How to help Indian students raise their levels of achievement is less well established.

This article reviews not only the research on the science and mathematics achievement of native Americans but also recommends three courses of action that teachers of Indian students can take to help students improve their academic preparation. Because the problem of low achievement in the sciences is also a national concern, this article may have implications beyond the population it targets.

The Need

The average achievement levels of native American students in science and mathematics are extremely low compared with students of other ethnic groups. In New Mexico, a tenth grade computation proficiency examination found only 21 percent of native American students scoring 65 percent correct or higher, while 27 percent of black students, 41 percent of Hispanic students, and 72 percent of white students scored at or above that level. In problem solving, the proportion of native American students achieving 65 percent or higher was nearly equal to black students (57 and 58 percent respectively), however, both were still much lower than Hispanic or white students (79 and 94 percent respectively) [18].

The California Achievement Tests' average mathematics score of Choctaw students attending four junior high schools in Mississippi was in the 17.8 percentile in 1977, the 22.3 percentile in 1978, and the 26.3 percentile in 1979 [2]. Although these scores indicate improvement, the average percentiles of Choctaw twelfth graders during that time frame was 17.5 in 1977, 25.8 in 1978, and 28.1 in 1979 [2].

This Mississippi study confirms a conclusion reached by the Senate Subcommittee on Indian Education ten years earlier: the native American child typically falls further and further behind national norms as he or she progresses through school [20]. This phenomenon has been designated as "progressive retardation." [5] The third grade Choctaw students in Brod's Mississippi study scored a 2.9 grade equivalency in mathematics in 1979. The progressive retardation apparent in his results

PHOTO BY OWEN SEUMPTEWA

shows sixth graders that year at 4.9 and twelfth graders at 8.1. The longer they were in school, the further they fell behind [2].

Bradley, in a paper on native American mathematics education, concludes:

> The evidence shows native Americans are not obtaining sufficient competence in mathematics to study the higher level mathematics courses in high school, to take the calculus sequence, and to enter mathematics-related careers. All the recruitment efforts of Indian communities and reservations, together with high technology industries to employ Indian people in careers requiring mathematics or statistics, will find little success as long as Indian students avoid mathematics and thus limit their choices in the job market. [1]

Green suggests the problem of native Americans' lack of achievement in science and mathematics is one of emotional or psychological origin. She interviewed Indian college students, educators, counselors, teachers, program directors, and many non-Indian educators and advisors for the AAAS Project on Native American in Science [8]. The purpose of her study was to find the barriers obstructing entry of native American students in the sciences and mathematics. Smith described several studies indicating that native American students living on reservations in Arizona have a significantly different understanding of words used in science and mathematics than either white or native American students from urban settings or urban children who speak only English at home: "Although the investigations are not conclusive, it does seem reasonable to suggest there are differences that exist, they are measurable, and they are culturally induced." [21] There are numerous hypotheses suggesting the causes and possible solutions to the problem of science and mathematics underachievement by native American students. Here, we discuss some of the more promising areas of investigation along with several hypotheses that need further research.

Even students performing well in science and mathematics often attribute their success to such external factors as luck or chance that may or may not be repeated. But these same students may attribute their failure so such internal factors as lack of ability.

Expectations

Participants attending the Conference on Mathematics in Indian Education in Albuquerque in 1978 "felt the factor most important in keeping Indian students from obtaining a good mathematics education is the prevalent feeling among teachers, counselors, and administrators that a more-than-rudimentary mathematics competence is "beyond and/or irrelevant to Indian needs." [10] Green, Brown, and Long, and Nash demonstrated effects on children's classroom and test performance resulting from teachers' expectations [10,14,15]. And, according to Ortiz-Franco, the influence of parents' expectations can be so great that their belief in their child's ability can predict course taking [17]. Many native American students report being counseled away from science and mathematics because it is perceived as too difficult for them or as unnecessary for their futures [8].

Student expectations are also significantly related to their ability to perform [7]. Thus, even students performing well in science and mathematics often attribute their success to such external factors as luck or chance that may or may not be repeated. But these same students may attribute their failures to such internal factors as lack of ability. The student locus of control for success is often different for success and failure [6]. Further research is needed to identify successful techniques for convincing native American students, their parents, teachers, and counselors that science and mathematics are not only important to these students' futures, but also that they do have the ability to perform in these fields.

Culturally Sensitive Programs

Participants at the Conference on Mathematics in American Indian Education considered culturally-based education to be especially appropriate for math and science [10]. Johnson recommends introducing science and math concepts via materials and examples from both the dominant culture and the specific ethnic culture of the students within the classroom [11]. He advocates encouraging students to interpret abstractions in the form of story situations that they contrive, using foods, occupations, places, and events from their environment. Green, *et al.* recommend looking at how the culture uses science and mathematics concepts (for example, astronomy, cooking, or crafts) and including these in the curriculum [10].

It is, however, difficult to generalize about what makes a successful program. There is no single native American culture upon which to base a science or mathematics curriculum. Rather, there are over 400 native American tribes, each with a different culture [12]. Because culturally based materials for one tribe may not be appropriate for students from other tribes, curriculum developers may need to look primarily at those universally accepted activities that are culturally sensitive for numerous tribes. Local classroom teachers can prepare science and math lessons with materials and examples appropriate for that specific tribe. For example, members of most tribes play games of chance; a unit on probability could be developed using examples from these games while encouraging teachers to include the game played by the specific tribe or tribes represented in the classroom.

Improving Pedagogy

Green's conclusions are bleak:

> I have concluded that, in general, the public school's preparation of most students in science and math is sadly wanting; that it is doubly wanting when it comes to answering the needs of minority students; that little attempt is made to address specific minority characteristics or needs when attempts are made to remedy the problem; and that Indian students who do not share in the fairly respectable public education afforded many of the students in my sample may be in worse shape than everyone else. [8]

Her concerns were echoed by participants of the Conference on Mathematics in American Indian Education. "All concurred that a basic science and mathematics education, as it is customarily taught, may be inappropriate for the needs of most people, and most certainly for Indians on reservations." [10] The Ford Foundation also agreed that the way science is taught to all student requires radical reform [13]. The results of the second National Assessment of Educational Progress

(NAEP) in math tend to support the same conclusions. Only 39 percent of nine-year-olds sampled believed they usually understood "what we are talking about in mathematics." [4]

Scott looked at the differences in performance in math between Pueblo Indian and white students entering a teacher training program [19]. The subjects of this survey were matched on overall math scores. The white students in this sample scored higher in arithmetic, while Pueblos scored higher in measurement. Scott noted that the arithmetic items were all highly symbolic and could be done with rote manipulation, devoid of any real world context, while the measurement items were application problems that related to real world experiences. He suggests that this may provide a useful teaching hint in providing basic math instruction to Pueblo groups—namely, that skill development be related to real-world situations, such as those involving measurement.

Competence in science and math is one of the elements needed to successfully negotiate a technological society. It is apparent from these studies, as well as investigations in more than a dozen Indian communities, that students have relatively low science and math achievement and high levels of science and math avoidance. The limited amount of research and preponderance of opinion indicate that programs that are culturally sensitive have greater success with Indian students than do traditional programs. The data indicate that even when native American students are successful in these fields, they still demonstrate avoidance. What we need is a program to both strengthen the skills of staff and convince parents, teachers, and counselors that native American students are capable of success in science and math, and further, that science and math are essential to success.

Teacher Education and On-Site Needs

Indian communities have a long history of problems in selecting, recruiting, and training teachers. Not only does the isolation of many reservation schools hamper recruitment, but most Indian communities cannot offer the lifestyle that many teachers expect. Therefore, these schools tend to be characterized by high teacher turnover.

Another factor contributing to this turnover is that a high percentage of first-year teachers are ill-prepared to meet the specific needs of native American students. These new teachers frequently arrive at their teaching sites with little or no targeted preparation in teaching Indian students. The teaching situations they encounter are often similar to what they might expect if they were teaching in a foreign country. Indeed, the native American classroom *is* foreign to the novice educator, especially a first-year teacher recruited from an urban college setting in a different section of the country. Thus, that teacher spends much time the first year trying to understand his or her new students and why they don't respond like those taught during the student teaching experience. "Why don't the students like me?" or "My students are too quiet and don't seem to understand the lessons" are common complaints from new teachers of Indian students.

These teachers soon learn that differences do exist and that they must use new approaches—not necessarily learned in their teacher training—if these students are to respond positively to instruction. Some estimate that, at a minimum, the first two years are devoted to classroom experimentation in an effort to discover what works. Of course, if they don't meet with quick success, many teachers leave at the end of the first year. More persistent teachers may stay longer, but, as a result of the multiple educational problems present in many Indian schools, they too, may leave eventually. These vacancies leave space for new teachers, and the high-turnover, low-success cycle continues.

Even those teachers who eventually discover and develop teaching strategies to which Indian students respond positively face other problems that work against a long-term commitment. Reservation locations often do not offer a permanent lifestyle or are not conducive to non-Indians establishing community roots. For example, non-Indians frequently cannot purchase land or make capital investments, thus making a long-term association impracticable.

Reservation locations often do not offer a permanent lifestyle or are not conducive to non-Indians establishing roots. Frequently, non-Indians cannot purchase land or make capital investments, thus making a long-term association impracticable.

The ideal solution, of course, would be more Indian teachers in math and science, but they are such a rarity in most Indian communities that this cannot be a real goal in the short run. In the Native American Science Education Association's (NASEA) Indian Science Teacher Education Program (ISTEP), NASEA staff visited 147 schools in Arizona and New Mexico and found only 2 certified secondary native American science or math teachers. Those native Americans who do choose careers in math and science tend not to enter teaching as a profession but instead select careers in medicine, higher education, or research, which provide higher salaries and greater visibility. Over the past twenty years, there has been a great deal of emphasis (with some success) on encouraging native Americans to enter these fields. But efforts to encourage them to select teaching careers in math and science have been virtually nonexistent. Many teacher education programs that highlight Indian education have been phased out of our universities.

Certified native American math and science teachers are desperately needed in these schools in order to provide the best possible learning environment. These teachers will be more sensitive to the connections between science and culture and can provide models for success in both. This goal must be immediate and ongoing, the object of our present attention and future objectives, but we must also address the current classroom situation. It is imperative that local school systems provide additional culture training for their new staffs.

Such orientations should include an introduction to the community and its expectations. Teachers should be made aware of cultural taboos before being put in the classroom. Teaching methods that work well with native American students should be included in the instruction. New teachers cannot expect instruction that works well in an urban setting to be culturally acceptable or effective in a reservation classroom. Such orientations can anticipate some of the cultural shock and frustrations new teachers experience by giving them both understanding and realistic strategies to increase their effectiveness in teaching Indian students.

Teachers should be aware of cultural taboos. They cannot expect instruction that works well in an urban setting to be culturally acceptable or effective in a reservation classroom.

NASEA's work has highlighted these observations and further shown that in the various types of schools serving Indian students:

- an extraordinarily high rate of teacher turnover exists in many reservation schools; (Thirty-three percent or higher in some subject areas in not unusual.)
- many teachers at the junior high level and above are not adequately prepared to teach the sciences, and virtually none of these is native American; and
- geographic and cultural isolation, compounded by poor recruitment practices and orientation efforts have contributed to staffing difficulties in reservation schools.

Complications

Within the last decade, native Americans have assumed greater control over their schools. As they have gained responsibility for school adminstration, they have had to grapple with educational priorities, the development of educational philosophies, and assessments of their own real needs. In nearly every case, this process has barely begun. It has been hampered by conflicting educational structures and the pressure of operating these systems in, at best, difficult circumstances. Faced with enormous need in local communities, education is seldom a top priority. Distorted by the fads of federal finances and the inconsistent efforts of constantly changing local educational personnel, progress in attaining educational objectives is slow and uncertain. The net loser is the Indian student. Local education efforts have too frequently become a well-meaning system that fails to adequately prepare students for success in higher education and loses nearly half its enrollment before secondary matriculation is complete. Each school is virtually an island with infrequent and usually only informal contact at the staff level with other schools in the local community.

This discouraging description is very clearly evident in the sciences (broadly defined to include math and social studies, as well as the physical, chemical, and biological sciences). Indian students have found this the most difficult area of the curriculum.

NASEA has identified two strategies to increase the size of the scientific/mathematical pool: first, to strengthen the quality of education available to native Americans before and during high school, through our precollege priority, with a special emphasis on teachers; second, to develop strategies to decrease attrition from the pool with special attention on the curriculum and its sensitivity to the unique needs of native American students.

This second strategy has special relevance to higher education. Statistics show that native American students drop out at higher rates at this level of education than any other (48–52 percent precollege, 60–86 percent higher education) [16]. This data highlights the special problems these students have in making the transition from high school to college and the need for special support systems to help them persist in the pursuit of their educational objectives. Many schools clearly have not identified effective mechanisms to meet these needs.

Gary G. Allen (above) is currently a development consultant to the American Indian Heritage Foundation, Falls Church, VA 22044 and was formerly executive director of the Native American Science Education Association, Washington D.C. 20005. Owen Seumptewa (below) is director of the Indian Science Teacher Education Program, S.W., Flagstaff, AZ 86005.

NASEA has come to believe that if native Americans are to assume key scientific and technical positions within their communities, there must be an overall improvement in the quality of precollege science and mathematics education available to native Americans and an increase in the number of students seeking degrees in science and engineering. Presently, Indian students are not prepared to successfully complete standard mathematics, science, and engineering programs in college. Their preparation is such that only extraordinary and expensive efforts have been able to assist in the development of the skills Indian students need to enter and ultimately complete professional studies. □

References

1. Bradley, C. "The State of the Art of Native American Mathematics Education." In *Handbook for Conducting Equity Activities in Mathematics Education,* H.N. Cheek, ed. Reston, Va.: NCTM, 1983.
2. Brod, R.L. *Choctaw Education.* Box Elder, Mont.: LPS & Associates, December 1979.
3. ———, and M.J. Brod. *Educational Needs of Colville Confederated Tribes.* Box Elder, Mont.: LPS & Associates, August 1981.
4. Carpenter, T.P., M.K. Corbitt, H. Kepner, M.M. Lindquist, and R.E. Reys. "Students Effective Responses to Mathematics: Results and Implications from National Assessment." *Arithmetic Teacher* 28(2):34–37, 52–53; October 1980.
5. Coombs, L.M. *The Educational Disadvantage of the American Indian Student.* Las Cruces, N.M.: New Mexico State University, 1970.
6. Crandall, V.C., W. Katkovsky, and V.J. Crandall. " Children's Beliefs in Their Own Control of Reinforcements in Intellectual–Academic Achievement Situations." *Child Development* 36:91–109; 1965.
7. Fennema, E. "Girls and Mathematics: The State of the Art." In *Handbook for Conducting Equity Activities,* H.N. Cheek, ed. Reston, Va.: NCTM, 1983.
8. Green, R. "Math Avoidance: A Barrier to American Indian Science Education and Science Careers." *BIA Educational Research Bulletin* 6(3):1–8; September 1978.
9. Green, R. "AAAS News: Math Called Key to Indian Self-Determination." *Science 201* August 4, 1978.
10. Green, R., J.W. Brown, and R. Long. *Report and Recommendations: Conference on Mathematics in American Indian Education.* Educational Foundation of America and American Association for the Advancement of Science. Washington, D.C.: February 1978.
11. Johnson, W.N. *Teaching Mathematics in a Multicultural Setting: Some Considerations When Teachers and Students Are of Differing Cultural Backgrounds.* Murray, Ky.: Murray State University, 1975. (ERIC Document # ED 183 414).
12. Keshena, R. "Relevancy of Tribal Interests and Tribal Diversity in Determining the Educational Needs of American Indians." In *Report of the Conference on Educational and Occupational Needs of American Indian Women.* National Institute of Education. Washington, D.C.: October, 1980.
13. *Minorities and Mathematics.* Ford Foundation. New York: 1981.
14. Nash, R. *Classrooms Observed: The Teacher's Perception and the Pupil's Performance.* Boston: Routledge & Kegan Paul, 1973.
15. Nash, R. *Teacher Expectations and Pupil Learning.* Boston: Routledge & Kegan Paul, 1976.
16. *Native Americans in Science,* unpublished monograph developed for Native American Science Education Association. Washington, D.C.: 1987.
17. Ortiz-Franco, L. *Suggestions for Increasing the Participation of Minorities in Scientific Research.* Washington, D.C.: National Institute of Education, April, 1981.
18. *A Planning Grant Proposal for a Comprehensive Mathematics Program in Northern New Mexico.* Southwest Resource Center for Science and Engineering. Albuquerque: 1981.
19. Scott, P.B. "Mathematics Achievement Test Score Comparison of American Indian and Anglo Students Entering and Elementary Teacher Training Program." *Journal of American Indian Education* 22(3);17–19; 1983.
20. Senate Subcommittee on Indian Education. *Indian Education: A National Tragedy, A National Challenge.* Washington, D.C.: U.S. Government Printing Office, 1969.
21. Smith, L. *Mathematics Education in an American Indian Culture.* Unpublished manuscript. Copy available from author, Mathematics Department, Arizona State University, Tempe, Ariz. 85287.

DEBORAH J. TIPPINS and
NANCY FICHTMAN DANA

ASSESSING FOR INDIVIDUAL DIFFERENCES

Culturally Relevant Alternative Assessment

Because experience is embedded within culture, learning should be considered in a cultural context. Similarly, assessment involves the representation of knowledge and a judgment concerning the viability of that knowledge. Thus, assessment should be context dependent; reflect the nature of the subject matter; and address the unique cultural aspects of class, school, and community among culturally diverse populations.

Why culturally relevant assessment?

Broad movements for reform and change in science and mathematics—such as Project 2061; the Scope, Sequence, and Coordination Project; and Curriculum, Evaluation, and Professional Teaching Standards for School Mathematics—must be influenced by the development of more culturally responsive means of assessing student learning in science and mathematics. The need for culturally relevant assessment reflects the diversity of our society, where students of color are expected to comprise 33 percent of public school enrollment by the year 2000.

Assessment in the next decade must challenge old assumptions. The inadequacy of standardized testing as a sole measure of what students know has been compounded because these tests too often portray an inaccurate picture of minority students' capabilities. As the Quality of Education for Minorities Report (1990) points out, test scores alone are poor measures of student potential.[1] Such measures fail to consider interpersonal skills, language abilities, and related talents that students will need in the real world. Beliefs and knowledge about culturally diverse groups may also have served to limit perspectives, ultimately contributing to reduced opportunities in fields such as science. Culturally relevant alternative assessment is needed to improve educational options for students from diverse backgrounds.

A critical factor in culturally relevant assessment is the realization that alternative assessment in multicultural populations will be different from other forms of assessment. Because all students bring with them a unique set of experiences, the developers of alternative forms of assessment must take these experiences into consideration. This is critical if one views improving learning as the primary purpose of assessment.

Assessment for all

The following alternative assessment strategies enable students in science to "show what they know" regardless of their cultural background. Furthermore, many of these strategies celebrate cultural diversity, for it is through alternative assessment strategies that both students and teachers from diverse backgrounds may learn to appreciate the uniqueness as well as the universality of their particular culture.

Concept mapping

Concept maps are constructed by selecting and writing major concepts and ideas in a circle or oval, and then joining related concepts with lines and connecting verbs that explain the relationships between concepts. Concept maps become tools for negotiating meanings between students when a culturally diverse group of two or three students must share, discuss, negotiate, and agree upon meanings in order to create a concept map. Novak and Gowin suggest that bilingual students may present foreign words that label the same events or objects.[2] By doing this, science students learn that language does not make a concept, but rather language serves as the label for a concept. A pre and post concept map (See Figure 1) can be one culturally relevant tool used to assess student learning in science.

Cooperative learning and group assessment

The common theme in all cooperative-learning, group-assessment strategies is that the grade earned fosters positive interdependence—how students work with one another is a prime factor in the grade received. Thus, students engage in dialogue to construct and negotiate a shared meaning of their science learning, and group members benefit by helping one another. In the process, students come to understand and appreciate cultural differences. Research on cooperative learning involving different ethnic groups, handicapped and nonhandicapped students, and male and female middle school students indicates that "cooperative learning experiences, compared to competitive and individualistic ones, promote more positive attitudes toward members of a different ethnic group or sex and handicapped peers."[3] Thus, cooperative learning not only offers an alternative to traditional, individualized, competitive assessment practices; but promotes attitudinal changes between culturally diverse students as well.

Figure 1. Student Pre and Post Concept Map

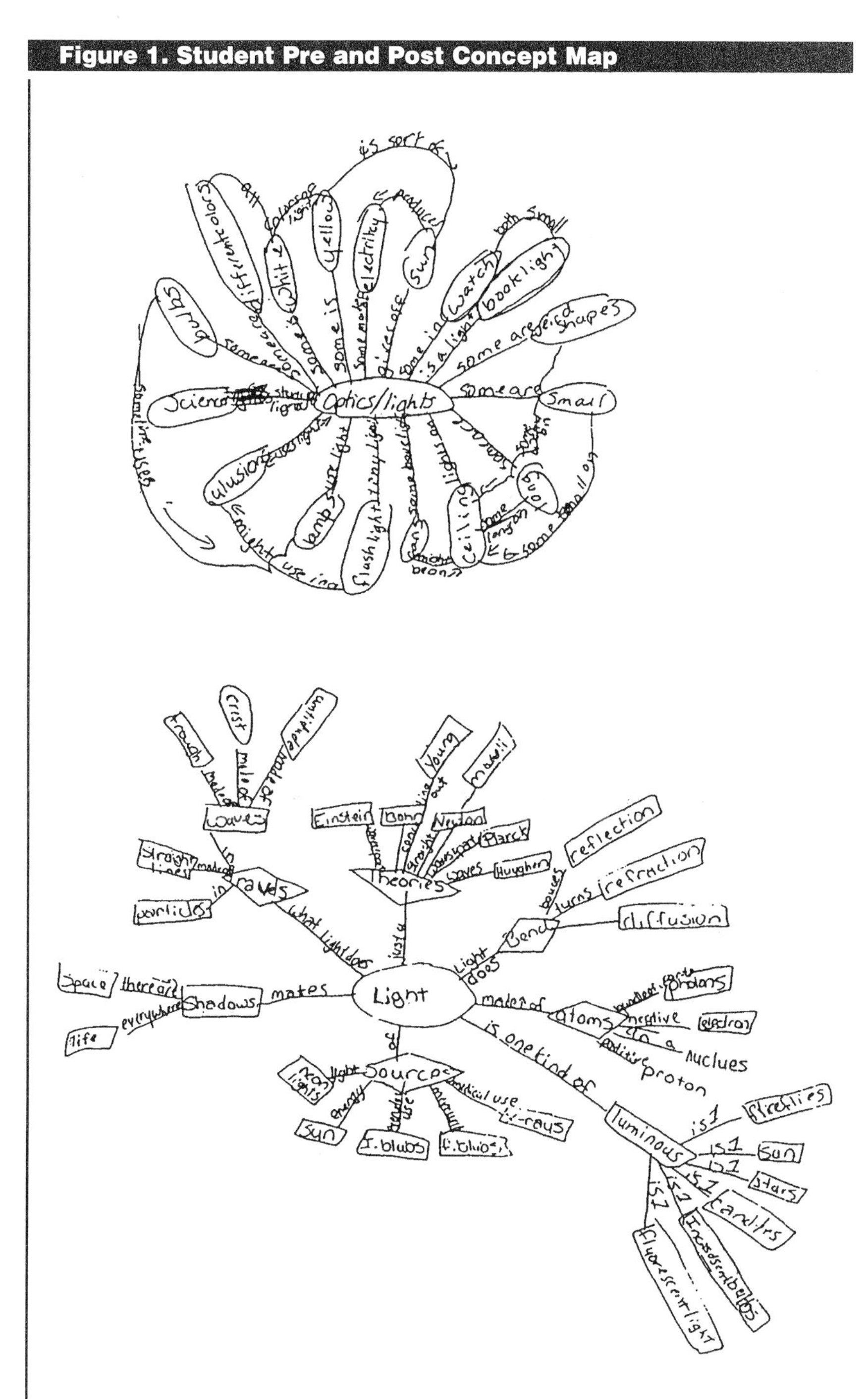

Journaling

Journaling becomes an alternative approach to traditional science testing when it is used for assessment purposes. Journal writing encourages students to connect science to experiences in their own lives and it helps students find connections between experience and theory.[4] Journaling is also culturally relevant in that it is a personal process in which grammar and punctuation are unimportant, thus not placing ethnic students at a disadvantage due to linguistic differences in grammar, lexicon, and style of dialect. When students keep journals of their science learning, they articulate their thought processes. When students share journal entries with one another, they may come to appreciate linguistic differences and the expressive power of language.

For assessment purposes, a unit of study may begin with students writing an entry on what they know about a particular scientific concept. Throughout the unit, students continue to write entries regarding their learning and how it relates to their life experiences.

Dialogue journals and *roving journals* are two additional ways of using journals for assessment. They enable students to communicate with each other about their learning. Dialogue journals place students in two-way conversations,

Figure 2.

Date : Tuesday, September 3
Topic : Reflection on the first week's activity

I learned that Mr. Hook is a cool dude who assigns homework most of the time. Also not to sterotype the color or sex of a scientist

It has changed my view of the world by, All scientist are not white, and all are not male

What I will do with this information is when I draw scientist I will draw minorities, and of the oppisite sex. And spread the news to my friends.

Good

My mom is a chemist and she is Indian.

So she. That's good to know.

focusing on the process of making sense in learning science. With dialogue journals, students make sense of their individual involvement with key science concepts by "talking with a friend." Students exchange journals on a regular basis, providing feedback and comments that facilitate further reflection. Roving journals are a powerful vehicle for helping students search for connections between their prior knowledge and new information. Roving journals focus on a particular science concept or activity. The journal is passed around the classroom and each student writes about the theories he or she used to make sense of the concept. As different students write explanations, students benefit from the ideas of the students before them.

When using any type of journals, it is important for students to consider how their thinking develops and reflect upon learning experiences both historically and in a cultural context. When students are allowed to communicate through journals, they become personally involved in learning and ultimately become participants in the assessment process. Figure 2 illustrates how teachers can use dialogue journals to assess student understanding of science concepts.

Oral interviews

Oral interviews provide an alternative assessment strategy that encourages students' self-confidence in posing questions in their own language in the context of their own experience. Teachers and students alike need opportunities to share their "stories," for in doing so, they demonstrate their individual questions and perspectives, which are essential in weaving the threads necessary to make sense of experience. Some topics around which oral interviews might initially be structured include job interviews and related accomplishments, interviews concerning controversial issues (including both biased and nonbiased accounts), and science-related oral histories.

Portfolios

The portfolio should provide a developmental record of growth in conceptual understanding for both teachers and students. The use of portfolios in assessment reflects a fundamental change from traditional assessment practices in many ways: The development of portfolios allows teachers and students to work and learn together; provides opportunities for reflection and self-assessment; helps redefine traditional student and teacher roles in relation to the science curriculum; emphasizes the culture in which teaching and learning occurs; and empowers both students and teachers with respect to science learning.

Many of the assessment tools already mentioned can be introduced into the assessment portfolio: concept maps, journals, oral interviews, and cooperative group assessment. Using multiple sources to profile student growth can help insure equitable treatment of culturally diverse students. But portfolios cannot be suddenly introduced into a science classroom without corresponding changes in the way we think about knowledge construction in relation to science teaching and learning. This calls for radically transformed science classrooms and new roles for teachers and learners.

Conclusions

In *Multiethnic Education*, Banks calls for the development and use of novel assessment practices that reflect various ethnic cultures.[5] It is important, however, to heed Banks' warning that alternative assessment strategies will do little good unless educators implement curricular and instructional practices that are also multiethnic and multiracial. The task of "fitting the school to the student" is difficult when schools define learning as "mastery" of isolated bits of data and marginalize conceptual knowledge and the related processes of problem solving and problem detecting. The strategies we present here cannot substitute for a multiethnic and multiracial science curriculum and learning environment. Any discussion of assessment must be linked to understanding what we teach, how we teach, and why we teach. A multiethnic and multiracial science curriculum and learning environment, as well as culturally relevant assessment strategies, can enable teachers to better prepare their students for the multicultural society in which we live ■

References

1. Quality of Education for Minorities Report. (1990). *Education that Works: An Action Plan for the Education of Minorities*. Cambridge, MA: Massachusetts Institute of Technology.
2. Novak, J. D., and Gowin, D. (1984). *Learning How to Learn*. Cambridge, England: Cambridge University Press.
3. Johnson, D. W., and Johnson, R. T. (1978). Cooperative, Competitive, and Individualistic Learning. *Journal of Research and Development in Education, 12*(1), 3–15.
4. Grumbacher, J. (1987). How Writing Helps Physics Students Become Better Problem Solvers. In T. Fulwiler (Ed.), *The Journal Book* (pp. 323–329). Portsmouth, NH: Boynton/Cook Publishers.
5. Banks, J. (1988). *Multiethnic Education*. Newton, MA: Allyn and Bacon.

Deborah J. Tippins is an assistant professor of science education at the University of Georgia and **Nancy Fichtman Dana** is a visiting assistant professor at The Florida State University, Department of Childhood Education.

SCIENCE Across Cultures

Part I: African and Native American achievements

ART BY TOM TILTON

by Helaine Selin

Since the European scientific revolution of the seventeenth and eighteenth centuries, we have come to regard Western science as the only true scientific enterprise. We think of science as objective, truthful, progressive, and free of superstition and cultural limitations.

Astrology and Alchemy states, "It has been customary to treat as folly and superstition many...past attempts at knowledge of nature or at understanding her phrases and processes. It is not uncommon to casually stumble upon some excellent text in the physical sciences or even some erudite modern treatise in the philosophy of science, and read there that science began with Galileo, that the Greek or medieval scholars did not know the meaning of an experiment or of evidence, that people did not know to count the teeth of a horse...and many more such foibles."

The Eurocentric view has been challenged in many fields over the past few years and, in response, teachers have begun to restructure their curricula on all educational levels. Many offer classes on African American, Native American, Chinese, and Latin American history and literature.

Teachers have also found themselves under pressure from parents and school districts to incorporate multiethnic and multicultural studies into their classes. However, during this rewriting and rethinking, the scientific achievements of non-Western cultures are often overlooked. While we are quick to recognize the charm, importance, and superior ability involved in African painting or sculpture, Brazilian music, or Indonesian textiles, we still look skeptically on non-Western medical and agricultural practices.

African Science: Myth or Reality reports that "Western peoples have simply absorbed only those aspects of Egyptian, Roman, and Arab science that appeal to a people with a mechanistic frame of mind. What they do not understand they call magic."

In my book, *Science Across Cultures,* I state that: "If we define science and technology broadly as ways of observing, describing, explaining, predicting, and controlling events in the natural world, then it is obvious that each culture has its own characteristic science and technology. The people of different cultures have developed different ways of perceiving and interacting with the natural world, because they have lived in different environ-

> *Cultural attitudes and ideas have affected the development and application of science throughout history.*

ments and have sought solutions to different problems. Moveover, the distribution of science and technological information; the means of recording, transmitting and disseminating such information; and the social status of scientists and technologists have varied from one culture to the next."

Cultural attitudes and ideas have affected the development and application of science throughout history. Once they are introduced to the scientific traditions of other cultures, students will gain an understanding of the relationship between science, its technological applications, and the cultures that benefit from those applications.

In this article, I will highlight a number of scientific and medical accomplishments of other African and Native American cultures. The article in next month's issue will present information on similar achievements of Chinese and Islamic people.

The information presented is only a small sample of the available literature on this subject. There are many books, films, and videos that you can use in preparing curricula that include the scientific achievements of these and many other cultures. Should you wish to research these topics, my book, *Science Across Cultures,* lists over 800 books for scholars and laypeople on the scientific achievements of other cultures. A bibliography of about 100 titles suitable for the high school level and a videography and filmography are included here (see pages 43 and 44).

AFRICA

Although scholars set Egypt apart from other African nations for many years, Egypt is, of course, geographically a part of Africa; the Rift Valley that begins in Egypt extends to South Africa. And we now know that Egypt was racially a part of Africa as well; examination of papyri and paintings clearly reveals that many Egyptians were black Africans.

The remains of the earliest humans have been found in Africa and evidence of ancient African science also exists. Some of the well-known contributions of ancient African science include one of the first intensive agricultural schemes; metallurgy, including the mining and smelting of copper, practiced in Africa as far back as 4000 B.C.; and the system of hieroglyphic writing and the use of papyrus.

The science of architecture also reached new heights with the pyramids. They were amazing accomplishments both in terms of construction and the mathematical and astronomical knowledge necessary to build and situate them.

Between 3000 and 2500 B.C., a calendar and numeration system were developed and a carefully defined medical system was established under the guidance of Imhotep, an African physician and architect. This section will focus on these two aspects of African science.

MATHEMATICS AND THE CALENDAR

The Ishango bone, found in what is now Zaire, in Central Africa, bears markings that indicate it was probably used as a record of months and lunar phases. The bone has been dated from between 9000 and 6500 B.C.

The markings on the Ishango bone consist of sets of notches in three columns, arranged in distinct patterns. One column of four groups is composed of 11, 13, 17, and 19 notches, the prime numbers between 10 and 20. In another column, the groups consist of 11, 21, 19, and 9 notches. The pattern here may be 10+1, 20+1, 20-1, and 10-1. The third column is arranged in a way that suggests a doubling operation. One can conclude from this that the people who made these notations used a base ten number system, knew how to double, and were familiar with prime numbers.

Besides indicating a system of notation and counting, the Ishango bone provides evidence that African people were marking time 8000 years ago. According to *The Roots of Civilization,* when the marks are plotted against a lunar model, there is a "close tally between the groups of marks and the astronomical lunar periods."

Other ancient cultures counted time as well. The Egyptian calendar was the first to use a 365-day year. Environmental and astronomical indicators also helped the Egyptians establish their calendar. For example, they were aware that the Nile River normally began to flood when Sirius, the Dog Star, one of the brightest stars in the heavens, rose along with the Sun.

The Egyptians divided the year into three seasons of four months each: flooding, seeding, and harvesting. This lunar cycle based system would have been fine, except for the fact that the year is actually about five hours longer than 365 days. At first, the priests just added an extra month every few years. This adjustment in time proved too chaotic, and finally a civil calendar was created that was based on a solar year divided into 12 months of 30 days each, with an extra five days tacked on at the beginning. This simple, orderly calendar predated the Aztec calendar by about 3000 years.

SURGERY AND GYNECOLOGY

Egyptians were responsible for many medical innovations. In addition to developing an elaborate herbal tradition and many methods of clinical therapy, they devised a code of medical ethics, urging physicians to treat the poor, to take little for themselves, and to live well in order to be able to pass on their skills.

Most societies have tried to find ways of preventing, encouraging, and stopping birth. The Egyptians used plants for contraception whose scientific validity has been proven by modern biomedical methods 4000 years later.

These contraception practices are described as follows in *The Medical Skills of Ancient Egypt*: "The oldest of all the surviving medical papyri, the Kahun gynecological Papyrus, which was compiled around 1900 B.C. during the Middle Kingdom, lists several recipes for contraceptives to be inserted into the vagina. They included pessaries made of honey with a pinch of natron, of crocodile dung in sour milk, or sour milk alone, or of acacia gum. Recently the latter has been recognized as spermatocidal in the presence of vaginal lactic acid, and it is conceivable that sour milk, too, might be an effective spermatocide."

Celery was also used as a contraceptive. Mixed with equal parts of oil and sweet beer, boiled, and taken for four mornings, it was said to prevent conception. A large dose of this mixture was said to induce abortion. Also used were dates, onions, and the Fruit-of-the-Acanthus, crushed in a vessel with honey, sprinkled on a cloth, and applied to the vulva. According to sources, it ensured abortion either in the first, second, or third trimester.

The Egyptians were also among the first to practice surgery. While many societies used surgical techniques for ornament (ear piercing) and medicine (bone setting), *The Medical Skills of Ancient Egypt* acknowledges that Egyptians were among the first to circumcise men. "It was a ritual performed by priests, probably on large groups of adolescents or young men (but not infants). During the Old Kingdom only royalty, nobility and priests were circumcised routinely. In later centuries the rite may have become obligatory for all pubertal males, perhaps as a precondition of marriage, but it may have

The Egyptians used plants for contraception whose scientific validity has been proven by modern biomedical methods.

been optional or even unavailable for some young men. Regardless of the extent to which circumcision was practiced, it seems to have grown out of the priests' concern for bodily cleanliness and, hence, purity...The world's oldest portrayal of circumcision—in fact, the first known picture of any surgical technique—was carved on the wall of a Dynasty VI tomb about 250 years after Imhotep's death."

NATIVE AMERICANS

As a result of recent archaeological findings, many of our ideas about Native American culture and science have changed. The image of nomadic hunter-gatherers has been replaced by that of a culture that resourcefully adapted to environmental conditions, carried on extensive trade and intertribal relations, and developed considerable artistic and technical skills.

There are many different Native American people, and what holds true for one tribe may not be for all tribes. However, for our purposes here, we will treat Native Americans as one, and presume that there were enough similarities in astronomical and agricultural practices to make this intellectually acceptable.

ASTRONOMY AND ARCHITECTURE

Like other early cultures, the Anasazi situated their dwellings and buildings in accordance with astronomical knowledge. In *The First Stargazers: An Introduction to the Origins of Astronomy*, this description is given: "The shape and layout were deliberately conceived to make maximum use of solar energy. The great semicircular shape open to the south is an effective

solar collector, with the high northern walls both reflecting the Sun's rays into the central plaza in winter and protecting it from the prevailing northwest winds."

Often buildings or boulders were assembled in a way that supplied information about the Sun's motion, in order to produce a calendar. Other sites were designed so that certain astronomical phenomena, such as equinoxes, were apparent by the movement of light through holes or windows. At certain times of the year, the light enters a specially placed window and reflects on an altar or other ceremonial spot. Marking of the divisions of the solar year by means of the midday Sun, rather than by the rising and setting of the sun, is evidence of an advanced astronomical system.

We think of ourselves as the most scientifically and technologically advanced people ever.

AGRICULTURE

We practice a very advanced form of agricultural technology, yet we find ourselves destroying the land and having to make use of more and more chemicals to fight off insects and keep the soil productive. We do this often at the expense of the workers who produce the food and the land that supplies it.

Native Americans practiced forms of agriculture that were harmonious with the land (even in arid or forested areas), and yet were still highly productive. The system was self-supporting and self-renewing, as the best cultivars were selected, recorded, and reused. Indigenous animals provided fertilizers. *Indian Givers: How the Indians of the Americas Transformed the World*, describes the traditional agricultural system of North and Central America as one that "centered on the small field called a *milpa,* which was not cultivated by plowing or planting in neat rows. The Indian farmer made a field of small mounds on which to plant the corn. In contrast to plowed rows, the small mound loses less soil to rain runoff and thus helps to stabilize the soil. White farmers in America adopted the practice, known as hilling, and followed it consistently from early colonial times until the 1930s. Since the United States abandoned hilling in favor of dense planting, erosion has increased remarkably."

Another area in which Native Americans showed agricultural superiority was in intercropping. We tend to practice monoculture—one field devoted to one crop. They created plots of mixed crops, usually beans, corn, and squash.

What may look chaotic to our eyes is in fact a wise and sensible polyculture system. The wide leaves of the corn plant protect the beans from the Sun; the hardy stalk serves as a stake on which the squash and bean vines grow. The vines provide good ground cover, reducing the need for weeding, helping to keep the soil moist, and protecting it from wind and water erosion. The beans are nitrogen fixers that encourage the growth of the corn and squash. We would have to provide cover and stakes for the beans, and fill the soil with artificially produced nitrogen and weed killers to achieve the same result.

Many of the plants cultivated by Native Americans have become staples in the world's diet: corn, beans, potatoes, maple syrup, and chocolate.

We think of ourselves as the most scientifically and technologically advanced people ever. In some ways we are. And yet, all our science is based on that which came before, often thousands of years before it was developed in the West. Some systems have expanded and improved on ancient science, and in some cases the ancients appear to have had a more humane and harmonious system that is equally efficient.

The world of science is vast and limitless, and every culture has produced its own science, which is a unique reflection of its world view and philosophy. To limit ourselves and our students to the study of one way of looking at medicine, mathematics, or astronomy, is to deny a world of knowledge that is rich and fascinating. As the Bantu proverb says, "He who never goes visiting thinks his mother is the only cook."

Helaine Selin is the science librarian at Hampshire College, Amherst, MA 01002.

REFERENCES

Bryan, C.P., trans. 1974. *Ancient Egyptian Medicine: the Papyrus Ebers.* Chicago: Ares.

Carlson, J.B., and W.J. Judge. 1987. *Astronomy and Ceremony in the Prehistoric Southwest.* Albuquerque: Maxwell Museum of Anthropology.

Cornell, J. 1981. *The First Stargazers: An Introduction to the Origins of Astronomy.* New York: Scribner.

Estes, J.W. 1989. *The Medical Skills of Ancient Egypt.* Canton, Mass.: Science History.

Manniche, L. 1989. *An Ancient Egyptian Herbal.* Austin: University of Texas Press, published in cooperation with British Museum Publications.

Marshack, A. 1972. *The Roots of Civilization.* New York: McGraw Hill.

Neugebauer, O. 1962. *The Exact Sciences in Antiquity.* New York: Harper.

Weatherford, J.M. 1988. *Indian Givers: How the Indians of the Americas Transformed the World.* New York: Crown.

Zaslavsky, C. 1973. *Africa Counts: Number and Pattern in African Culture.* Boston: Prindle, Weber & Schmidt.

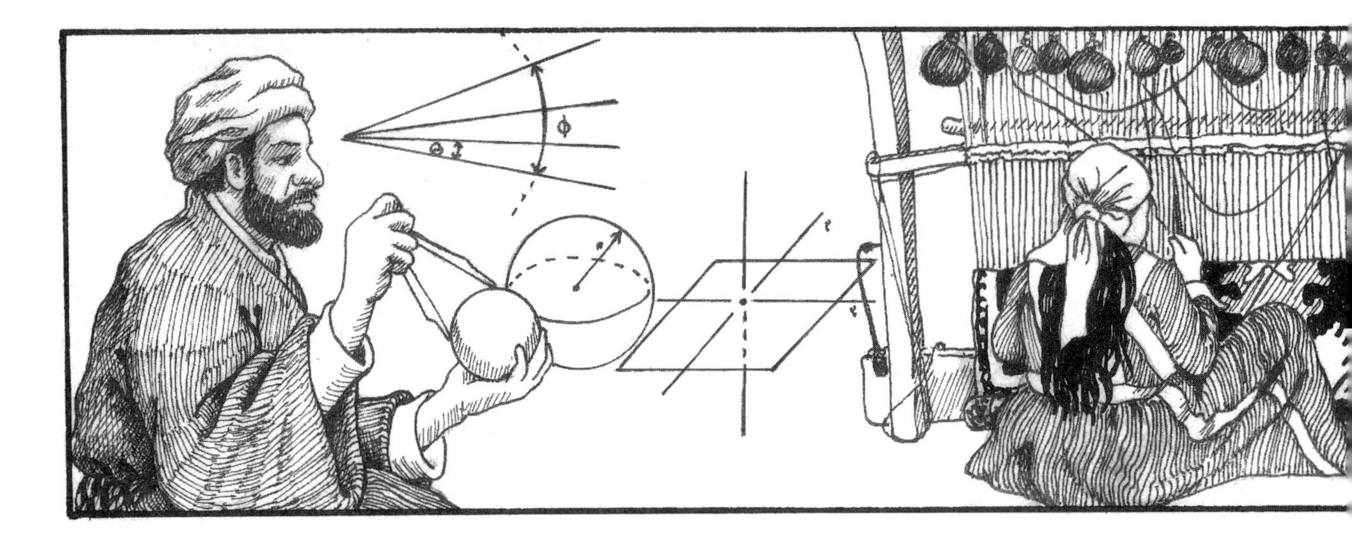

SCIENCE Across Cultures

Part II: Chinese and Islamic achievements

ART BY TOM TILTON

by Helaine Selin

In Part 1 of *Science Across Cultures* (*TST*, March 1993), I discussed some of the scientific and technical achievements of Africans and Native Americans. This month, I would like to continue the discussion of science in non-Western cultures by highlighting some of the contributions of China and the Islamic world.

CHINA

Science historians have focused more attention on Asia than on any other non-Western region. Asian science and technology was theoretically sophisticated and technologically advanced, and led the world for the 15 centuries prior to the Scientific Revolution in the West. China's contributions to science began as far back as the thirteenth century B.C., with the invention of lacquer, the first plastic. Europeans made this same "breakthrough" 3200 years later.

It is possible to make similar comparisons in many areas of science, from agriculture and astronomy to technology and medicine. Documents from the sixth century B.C. demonstrate that the Chinese believed in blood circulation 1800 years before Harvey made his "discovery."

The Chinese used the Mercator map projection in the tenth century A.D., 600 years before its use in the West. Many engineering feats, such as the suspension bridge, the chain-drive, and the essentials of the steam engine, preceded their so-called invention in Europe by over a thousand years.

In this article, I will focus on three very different inventions and achievements that illustrate China's impressive science and technology: the seismograph, methods of paper making, and Chinese medicine.

EARTHQUAKE DETECTION

China has always been beset by earthquakes. Often the quake would not just disrupt the Earth, but be followed by food riots or outbreaks of rebellion against the government. In *The Genius of China: 3000 Years of Science, Discovery, and Invention,* the need to detect earthquakes was recounted as follows: "The imperial government had every reason to want to know as soon as possible when there had been an earthquake in a distant province. First of all, it would mean that grain shipments would be interrupted, which was relevant since taxes were paid in grain. But it would also mean that both food aid and extra military forces would be needed in the afflicted area."

In 132 A.D., Chang Heng, the Astronomer-Royal, invented an instrument that could indicate when earthquakes had occurred. It was called an earthquake weathercock. The seismograph is described in *The Genius of China* as a "bronze vessel, rather like a winejar, six feet across, with a domed lid. The outer surface of the vessel was decorated with motifs of mountains, tortoises, birds, animals and antique writing. All around the vessel was a series of eight dragons' heads, equally spaced, holding bronze balls in their mouths. The balls would drop out if the dragons' mouths opened, or if pushed. Round the base of the vessel sat eight corresponding bronze toads, looking upwards, with their mouths wide open. They were positioned directly beneath the dragon mouths, ready to catch the falling balls. Obviously, a bronze ball dropping into a bronze toad would make a great deal of noise; people would be alerted by the resounding clang."

The making of paper is arguably China's greatest contribution to world culture.

The actual mechanism inside the vessel consisted of a weighted pendulum or bob that would tilt if there were an earth tremor, causing a ball to move along a slider and drop out of the dragon's mouth. The pendulum also released a hook, which fell and locked the other sliders in place, so that subsequent tremors could not send other balls in other directions, thus making it possible to detect both the time and the direction of the quake.

If the surviving reproductions of this early seismograph are credible copies, the machine was magnificently crafted. The first Western seismograph was designed in France in 1703, and the ones we use today were first developed in 1848.

PAPER MAKING

The making of paper is arguably China's greatest contribution to world culture, since paper made possible the preservation and dissemination of knowledge.

According to *Ancient China's Technology and Science,* "Paper was a brand-new type of writing material. Those used before had been tortoise shell, bone, metals, stones, bamboo slips, wooden tablets and silk . . . None of these materials, however, suited the need. Tortoise shell was scarce, metal and stone cumbersome, silk costly and bamboo slips and wooden tablets took up too much space."

The invention of paper is usually credited to Cai Lun, a eunuch who was Inspector of Public Works during the Han dynasty. In 105 A.D., he came up with an affordable and easy-to-make paper, created from vegetable fibers made from hemp rope ends, rags, and old fishing nets. The fibers were shredded and boiled, and then pounded into a soggy pulp. Then they were placed on a fine mesh screen in a water filled vat. After being strained, the fibers were left to dry, and the smooth dry fibers on the screen became paper. This process is virtually unchanged today, although the materials used have changed through the years. In fact, the Chinese have made paper from bamboo, rice and wheat straw, sandalwood, hibiscus, seaweed, floss silk, rattan, jute, flax, and ramie.

There is evidence that the Chinese made use of paper for many reasons besides writing. It was much coarser and stronger than paper today, and was used for clothing, lacquerware, and even military armor. Paper clothing was very warm and so impenetrable by cold winds that people complained that it allowed no circulation and was too hot to wear.

Paper was also used for wall decoration, and it is possible to trace the use of toilet paper in China back to the sixth century A.D. Paper for writing became the setting for Chinese artistry, and led to the development of calligraphy, water color painting, and block printing.

The movement of paper-making technology to the West was quite slow, traveling from India and West Asia in the seventh and eighth centuries. Eighth-century Arabs sold paper to the Europeans, but would not share the secret of its production. An equivalent paper making industry did not thrive in Europe until 1500 years after its invention in China.

MEDICINE

Although it is difficult to summarize Chinese medical achievements in this limited space, I will briefly discuss some of its basic tenets to give an idea of how extraordinary and complete a system it is. Traditional Chinese medicine is a completely different system from Western biomedicine, and evidence of its practice dates back to the first century A.D., with the *Yellow Emperor's Classic.* Acupuncture needles have been found buried in tombs from before the time of Christ.

Traditional Chinese medicine is holistic in its outlook, taking into account the season, weather, environment, family, diet, and situation of the unhealthy person, rather than looking at isolated symptoms. Diagnosis is performed by a series of four functions: looking, listening and smelling (interestingly, these two words are the same in the Chinese language), asking, and touching.

Another essential part of diagnosis is the patient's pulse reading. The physician takes the pulse in many places, looking for signs of disharmony, since the aim of medicine is to restore the body to harmoniousness. An elaborate system of pulse readings was devised, in which different pulses corresponded to different states of ill health. Illness is said to be caused by blockages in the flow of energy through a series of meridians and channels described in Chinese medical anatomy.

Illness is also described as an excess in either *yin* or *yang*, the two poles that are the basis for much Chinese philosophy. All things and all parts of things are either yin or yang, the way many European languages classify all nouns as either masculine or feminine. Too much of either is said to be the cause of distress, since the body is thrown into imbalance.

There are several means to a cure. The Chinese have one of the most advanced systems of pharmacology, using medications derived mostly from plants, but also from minerals and animals. They practiced diet therapy as far back as 200 A.D., and, in later years, recommended special diets for specific deficiency diseases. Some of their suggestions for dietary improvements are being incorporated by present-day physicians in their recommended nutritional therapies.

In addition, herbs are often applied to the skin, or burned and inhaled, in a process called moxibustion, which is a non-invasive, natural form of therapy. Moxibustion's partner in the therapeutic world is acupuncture, in which small needles are inserted into specific spots along the meridians and channels, in order to release the flow of energy and wellness. The Western world has been able to accept the efficacy of acupuncture as an anesthetic, when in fact it is primarily used to cure diseases.

The Arabs not only assimilated Greek science but made themselves masters of its methods and techniques.

If these examples sound questionably scientific, then you may wonder why they have survived all these years, and why Western doctors have gone to China to observe surgical procedures practiced without chemical anesthesia, and the curing of diseases from malaria to dysentery without using expensive medications produced in a laboratory. The most obvious answer is that they work.

Another answer is that these treatments are more accessible to a large rural population than Western medicine is. The fact remains that centuries before Europeans began the practice of experimental medicine, the Chinese had a fully developed and effective medical system that was harmonious with its environment, inexpensive, and successful. In a time of growing dissatisfaction with the cost and outcomes of Western medicine, traditional Chinese medicine has great appeal.

The Chinese contributed greatly to our scientific foundations, and there is a vast amount of literature on Chinese science and medicine that is readily available to teachers and their students.

ISLAMIC SCIENCE

In *The Genius of Arab Civilization: Source of Renaissance*, the author makes this observation: "The Arabs not only assimilated Greek science but also made themselves masters of its methods and techniques. Their role did not consist merely of handing over to Europe what they had earlier acquired from the ancients; rather, having digested what they learned from their predecessors, they were able to enrich it by new observations, new results, and new techniques."

In the ninth century A.D., the Muslims translated and disseminated many documents from other cultures and added their own, sparked by Islam's call to its believers to seek knowledge. Scientific knowledge was not limited to scholars, but was available to a large population. I will discuss two areas in which Islamic scientists excelled: mechanical engineering and astronomy.

MECHANICAL ENGINEERING

The evidence for remarkable inventive achievement in the Islamic world is contained in a book often translated as *The Book of Ingenious Devices* written in 850 A.D. by the Banu (Sons of) Musa bin Shakir. Most of these are trick vessels: pitchers that can pour hot, cold, or mixed water; a basin with a figure of a bull nearby that makes a thirsty sound when water is poured into the basin; a flask that discharges nothing when poured the first time, but works the second time; and a trough that replenishes itself when animals drink from it.

Islamic ingenuity was used not only for playful gadgetry, but for other feats of engineering and technology such as water-raising machines; water and mechanical clocks; roads and bridges; dams; shipbuilding; textile, paper and leather production; and agriculture and food technology. The Banu Musa were said to be the authors of several mathematics texts, including *On the Trisection of an Angle* and *On the Measurement of Plane and Spherical Figures.* They soon moved to a stage of research and innovation. The Banu Musa established an observatory at their own home where they carried out reliable astronomical observations.

ASTRONOMY

Islamic scientists made great advances in observational and mathematical astronomy. Many great observatories were established and funded, and Islamic astronomers were able to correct errors derived from calculations based on the belief that the Universe was geocentric (revolving around the Earth) rather than heliocentric (revolving around the Sun). One of the greatest achievements had to do with being able to calculate the direction of Mecca. Muslims are required to face Mecca and pray five times a day.

One of the greatest Islamic scientists, al-Biruni, was a political prisoner at one point in his life. While he was being transported to a prison in what is now Afghanistan, he attempted to determine the distance from and direction of Mecca. He was quite successful, and his operations led to the exploration of many related scientific areas such as terrestrial latitude and equinox observations.

According to *Islamic Science: An Illustrated Study*, in 1007 Ibn Yunus composed "a masterpiece of observational astronomy in which many constants have been measured anew and in which extensive use is made of trigonometry for the solution of astronomical problems. Ibn Yunus was also the first person to make a serious study of the oscillatory motion of a pendulum, which finally led to the invention of the mechanical clock."

The Muslims were also superior crafters of astronomical instruments. Their skill in the construction of astrolabes, instruments for measuring angles in the changing position of the stars and thereby charting their courses, was enormous.

It should also be mentioned that astronomy was often employed in conjunction with a sister science, astrology. In fact, in Arabic, the two words could be used interchangeably. *Islamic Science: An Illustrated Study* reports that "Although there were some authorities who accepted astronomy and condemned astrology, by and large the two intermingled and there was never in Islam the clear distinction that exists in the West today between astronomy considered as a science and astrology as a pseudo-science (with the embarrassing consequence that the supposedly pseudo-science seems to be attracting more Westerners than the science of astronomy itself in this supposedly most rational age of human history.)"

In my book, *Science Across Cultures*, in the introduction to the chapter on Islamic sciences, Ahmad Dallal, concludes that "The Arabic-writing cultures of the medieval Islamic societies supported scientific work more widely and more intensively than any of the societies which preceded them. Many of the findings of their scientists were transmitted to other regions and contributed directly to the further advance of the subject. Other Arabic discoveries were lost, to reappear elsewhere independently. In the centuries that followed the ninth century translation movement, Islamic scientists made significant contributions in all the fields they cultivated, and until the rise of modern science they were the best of their age."

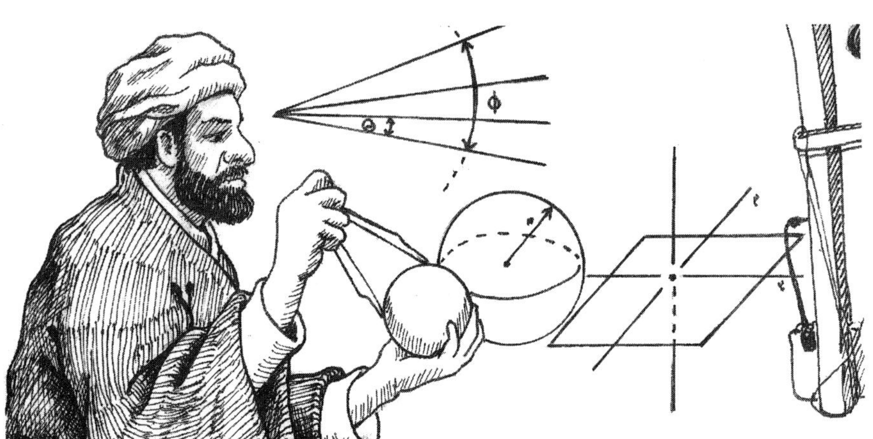

Literature on both Chinese and Islamic scientific and technological contributions are abundant and easy to incorporate into lessons on multi-cultural scientific achievements.George Sarton in his preface to *A History of Science*, wrote that "early Hindu science and Chinese science are generally left out...because they lack signification for us Western readers." It is my hope that we can prove him wrong.

Helaine Selin is the science librarian at Hampshire College, Amherst, MA 01002 and the editor of The Encyclopedia of Science Technology and Medicine in Non-Western Cultures.

REFERENCES

al-Hasan, A.Y. and D.R. Hill. 1986. *Islamic Technology: An Illustrated History.* Cambridge: Cambridge University Press.

Badeau, J.S and J.R. Hayes. 1983. *The Genius of Arab Civilization: Source of Renaissance.* Cambridge, Mass.: MIT Press.

Institute of the History of Natural Sciences, Chinese Academy of Sciences, comp. 1983. *Ancient China's Technology and Science.* Beijing: Foreign Languages Press.

Nasr, S.H. 1976. *Islamic Science: An Illustrated Study.* London: World of Islam Festival Publishing Company.

Selin, H. 1992. *Science Across Cultures: An Annotated Bibliography of Books on Non-Western Science, Technology, and Medicine.* New York, Garland Press.

Temple, R.K.G. 1986. *The Genius of China: 3,000 Years of Science, Discovery, and Invention.* New York: Simon and Schuster.

FOR FURTHER READING

Ackerknecht, E.H. 1973. *Therapeutics From the Primitives to the 20th Century.* New York: Hafner.

Akerblom, K. 1968. *Astronomy and Navigation in Polynesia and Micronesia: A Survey.* Stockholm: Etnografiska Museet.

al-Jazari, Ismail ibn al-Razzaz. 1974. *The Book of Knowledge of Ingenious Mechanical Devices.* Boston: Reidel.

Anderson, C.N. 1972. *The Fertile Crescent: Travels in the Footsteps of Ancient Science.* Fort Lauderdale, Fla.: Sylvester.

Ascher, M., and R. Ascher. 1981. *Code of the Quipu: A Study in Media, Mathematics, and Culture.* Ann Arbor, Mich.: University of Michigan Press.

Aveni, A.F. 1989. *Empires of Time: Calendars, Clocks, and Cultures.* New York: Basic Books.

Aveni, A.F. ed. 1990. *The Lines of Nazca.* Philadelphia: American Philosophical Society.

Badeau, J.S., and J.R. Hayes. 1983. *The Genius of Arab Civilization: Source of Renaissance.* Cambridge, Mass.: MIT Press.

Bastien, J.W. 1987. *Healers of the Andes: Kallawaya Herbalists and Their Medicinal Plants.* Salt Lake City: University of Utah Press.

Bose, D.M. ed. 1971. *A Concise History of Science in India.* New Delhi: Published for the National Commission for the Compilation of History of Sciences in India by the Indian National Science Academy.

Brecher, K., and M. Feirtag, eds. 1979. *Astronomy of the Ancients.* Cambridge, Mass.: MIT Press.

Breuer, H. 1972. *Columbus Was Chinese: Discoveries and Inventions of the Far East.* New York: Herder and Herder.

Bryan, C.P., trans. 1974. *Ancient Egyptian Medicine: The Papyrus Ebers.* Chicago: Ares.

Callicott, J.B., and R.T. Ames, eds. 1989. *Nature in Asian Traditions of Thought: Essays in Environmental Philosophy.* Albany: State University of New York Press.

Carlson, J.B., and W.J. Judge. 1987. *Astronomy and Ceremony in the Prehistoric Southwest.* Albuquerque: Maxwell Museum of Anthropology.

Chamberlain, V.D. 1982. *When Stars Came Down to Earth: Cosmology of the Skidi Pawnee Indians of North America.* Los Altos, Calif.: Ballena.

Chon, S. 1974. *Science and Technology in Korea: Traditional Instruments and Techniques.* Cambridge, Mass.: MIT Press.

Closs, M.P., ed. 1986. *Native American Mathematics.* Austin: University of Texas Press.

Cornell, J. 1981. *The First Stargazers: An Introduction to the Origins of Astronomy.* New York: Scribner.

Covell, A.C. 1986. *Folk Art and Magic: Sharnanism in Korea.* Seoul, Korea: Hollym Corp.

Dash, B., and M.M. Junius. 1983. *A Hand Book of Ayurveda.* New Delhi: Concept.

Day, C.L. 1967. *Quipus and Witches' Knots: The Role of the Knot in Primitive and Ancient Cultures.* Lawrence: University of Kansas Press.

de Rosny, E. 1985. *Healers in the Night.* Maryknoll, N.Y.: Orbis Books.

Densmore, F. 1974. *How Indians Use Wild Plants For Food, Medicine, and Crafts.* New York: Dover.

Dharampal, comp. 1971. *Indian Science and Technology in the Eighteenth Century: Some Contemporary European Accounts.* Delhi: Impex India.

Dommen, A.J. 1988. *Innovation in African Agriculture.* Boulder, Colo.: Westview.

Doshi, S., ed. 1985. *India and Greece, Connections and Parallels.* Bombay: Marg Publications.

Estes, J.W. 1989. *The Medical Skills of Ancient Egypt.* Canton, Mass.: Science History.

Ezeabasili, N. 1977. *African Science: Myth Or Reality?* New York: Vantage.

Felger, R.S., and M.B. Moser. 1985. *People of the Desert and Sea: Ethnobotany of the Seri Indians.* Tucson, Ariz.: University of Arizona Press.

Flegg, G. 1983. *Numbers: Their History and Meaning.* New York: Schocken Books.

Forbes, R.J. 1964–1972. *Studies in Ancient Technology.* Leiden: Brill.

Gelfand, M. 1985. *The Traditional Medical Practitioner in Zimbabwe: His Principles of Practice and Pharmacopeia.* Gweru, Zimbabwe: Mambo Press.

Ghalioungui, P. 1983. *The Physicians of Pharaonic Egypt.* Cairo: al-Ahram Center for Scientific Translations.

Gladwin, T. 1970. *East Is a Big Bird: Navigation and Logic on Puluwat Atoll.* Cambridge, Mass: Harvard University Press.

Goldstein, T. 1988. *Dawn of Modern Science.* Boston: Houghton Mifflin.

Graubard, M. 1953. *Astrology and Alchemy: Two Fossil Sciences.* New York: Philosophical Library.

Hadingham, E. 1988. *Lines to the Mountain Gods: Nazca and the Mysteries of Peru.* Norman: University of Oklahoma Press.

Hamarneh, S.K. 1983. *Health Sciences in Early Islam.* San Antonio: Noor Health Foundation and Zahra Publications.

Harley, J.B., and D. Woodward. 1987. *The History of Cartography.* Chicago: University of Chicago Press.

Hasan, A.Y., and D.R. Hill. 1986. *Islamic Technology: An Illustrated History.* Cambridge: Cambridge University Press.

Heiser, C.B., Jr. 1985. *Of Plants and People.* Norman: University of Oklahoma Press.

Heyn, B. 1987. *Ayurvedic Medicine: the Gentle Strength of Indian Healing.* Rochester, Vt.: Thorsons.

Ho, P.Y. 1985. *Li, Qi, and Shu: An Introduction to Science and Civilization in China.* Seattle: University of Washington Press.

Ho, Ping-ti. 1975. *The Cradle of the East: An Inquiry into the Indigenous Origins of Techniques and Ideas of Neolithic and Early Historic China, 5000–1000 B.C.*

Chicago: University of Chicago Press.
Hodges, H. 1970. *Technology in the Ancient World.* New York: Knopf.
Huard, P., and M. Wong. 1968. *Chinese Medicine.* New York: McGraw-Hill.
Ibn Butlan. 1976. *The Medieval Health Handbook: Tacuinum Sanitatis.* New York: Braziller.
Ifrah, G. 1985. *From One to Zero: A Universal History of Numbers.* New York: Viking.
Institute of the History of Natural Sciences, Chinese Academy of Sciences, comp. 1983. *Ancient China's Technology and Science.* Beijing: Foreign Languages Press.
Jaggi, O.P. 1969–1986. *History of Science and Technology in India.* Delhi: Atma Ram.
Kakar, S. 1982. *Shamans, Mystics and Doctors: A Psychological Inquiry into India and Its Healing Traditions.* New York: Knopf.
Kamal, H. 1975. *Encyclopaedia of Islamic Medicine, With a Greco-Roman Background.* Cairo: General Egyptian Book Organization.
Kaptchuk, T.J. 1983. *The Web That Has No Weaver: Understanding Chinese Medicine.* New York, N.Y.: Congdon & Weed.
Kaptchuk, T.J., and M. Croucher. 1987. *The Healing Arts: Exploring the Medical Ways of the World.* New York: Summit Books.
Khan, M.A.R. 1973. *Muslim Contribution to Science and Culture: A Brief Survey.* Lahore: Sri Muhammad Ashraf.
Kilpatrick, J.F., and A.G. Kilpatrick. 1970. *Notebook of a Cherokee Shaman.* Washington: Smithsonian Institution Press.
Krupp, E.C. 1983. *Echoes of the Ancient Skies: The Astronomy of Lost Civilizations.* New York: Harper & Row.
Kunzang, R.R.J. 1973. *Tibetan Medicine: Illustrated in Original Texts.* Berkeley: University of California Press.
Lewis, D. 1978. *The Voyaging Stars: Secrets of the Pacific Island Navigators.* New York: Norton.
Lewis, D. 1979. *We, the Navigators: The Ancient Art of Landfinding in the Pacific.* Honolulu: University Press of Hawaii.
Lewith, G.T. 1982. *Acupuncture: Its Place in Western Medical Science.* Wellingborough, England: Thorsons.
Li, Ch'iao-p'ing. 1948. The chemical arts of old China. Easton, Pa.: *Journal of Chemical Education.*
Majno, G. 1975. *The Healing Hand: Man and Wound in the Ancient World.* Cambridge, Mass.: Harvard University Press.
Manniche, L. 1989. *An Ancient Egyptian Herbal.* Austin: University of Texas Press, published in cooperation with British Museum Publications.
McNeil, I., ed. 1989. *An Encyclopaedia of the History of Technology.* New York: Routledge.
Merson, J. 1990. *The Genius That Was China: East and West in the Making of the Modern World.* New York: Overlook.
Nabhan, G.P. 1989. *Endunng Seeds: Native American and Wild Plant Conservation.* San Francisco: North Point Press.
Nasr, S.H. 1976. *Islamic Science: An Illustrated Study.* London: World of Islam Festival Publishing Company.
Needham, J., and C.A. Ronan. 1978. *The Shorter Science and Civilization in China.* New York: Cambridge University Press.
Nielsen, H. 1987. *Medicaments Used in the Treatment of Eye Diseases in Egypt, the Countries of the Near East, India and China in Antiquity.* Odense, Denmark: Odense University Press.
Pacey, A. 1990. *Technology in World Civilization: A Thousand-year History.* Cambridge, Mass.: MIT Press.
Palos, I. 1971. *The Chinese Art of Healing.* New York: Herder and Herder.
Reid, D.P. 1987. *Chinese Herbal Medicine.* Boston: Shambhala.
Rogers, S.L. 1985. *Primitive Surgery: Skills Before Science.* Springfield, Ill.: Thomas.
Ronan, C.A. 1973. *Lost Discoveries: The Forgotten Science of the Ancient World.* New York: McGraw-Hill.
Schultes, R.E. 1988. *Where the Gods Reign: Plants and Peoples of the Colombian Amazon.* Oracle, Ariz.: Synergetic Press.
Scully, V. 1970. *A Treasury of American Indian Herbs: Their Lore and Their Use For Food, Drugs, and Medicine.* New York: Crown.
Singer, C., E.J. Holmyard, and A.R. Hall., eds. 1954–1984. *A History of Technology.* Oxford: Clarendon.
Smith, D.E. 1958. *History of Mathematics.* New York: Dover.
Sung. Y. 1966. *Tien-kung kai-wu: Chinese Technology in the Seventeenth Century.* University Park: Pennsylvania State University.
Temple, R.K.G. 1986. *The Genius of China: 3,000 Years of Science, Discovery, and Invention.* New York: Simon and Schuster.
Thorwald, J. 1962. *Science and Secrets of Early Medicine: Egypt, Mesopotamia, India, China, Mexico, Peru.* New York: Harcourt, Brace & World.
Tillotson, A.K., and M.B. Bajracharya. 1987. *The Handbook of Ayurvedic Medicine: Science of Life.* Virginia Beach: Grunwald and Radcliff.
Tooley, R.V. 1970. *Maps and Mapmakers.* New York: Crown.
Turnbull, D. 1989. *Maps Are Territories, Science Is an Atlas: a Portfolio of Exhibits.* Victoria, Australia: Deakin University.
Van Sertima, I., ed. 1983. *Blacks in Science: Ancient and Modern.* New Brunswick, N.J.: Transaction.
Waheenee. 1987. *Buffalo Bird Woman's Garden.* St. Paul: Minnesota Historical Society.
Wang, X.T. ed., 1987. *An Illustrated History of Acupuncture and Moxibustion.* Beijing.
Watson, H., and D.W. Chambers. 1989. *Singing the Land, Signing the Land: A Portfolio of Exhibits.* Victoria, Australia: Deakin University.
Watson, W. 1971. *Cultural Frontiers in Ancient East Asia.* Edinburgh: Edinburgh University Press.
Weatherford, J.M. 1988. *Indian Givers: How the Indians of the Americas Transformed the World.* New York: Crown.
Zaslavsky, C. 1973. *Africa Counts: Number and Pattern in African Culture.* Boston: Prindle, Weber & Schmidt.

"If I have seen further than most men, it is because I stood on the shoulders of giants"

Isaac Newton

Art by Ruth Ketler

Not all of those giants were European

Arabic scientists in the Middle Ages

By Bruce Reichert

Few teachers are aware that non-Western scientists have made enormous contributions to scientific knowledge. In fact, an article in *Science Scope,* which offered the impressive idea of creating a science "time-line" in the classroom, went on to suggest that, ". . . if your wall space is limited, you may want to accordion-pleat the Middle Ages—between about 200 AD and 1300 AD—where there is plenty of empty space."[1]

During the Middle Ages, while Europe was experiencing a "dark" era and few scientific ideas were being presented, science and the pursuit of knowledge were *not* dormant. Incredible amounts of scientific knowledge and data were being gleaned, nurtured, expanded and stored in the Arab world, and some of this information would later stimulate Europe to its "Renaissance." In the seventeenth century Isaac Newton said, "If I have seen further than most men, it is because I stood on the shoulders of giants." As Newton well knew, not all of those giants were Europeans!

So, with the hope of preventing teachers from pleating a millennium of scientific history in an attempt to save wall space, I would like to share my knowledge of scientific "giants" from the Arab world.

Earth science and astronomy

Though the Islamic world never broke completely with Ptolemy's geocentric theory of the universe, Moslem scientists added to our understanding of astronomy by naming new stars, using astrolabes to ascertain the position of celestial bodies, recording celestial events, and creating a system to measure stellar magnitudes.

Moslem astronomers' interest in the heavens was directly related to religion, for astronomy allowed them to calculate and refine their lunar calendar, determine more accurately the beginning of the holy month of Ramadan, and locate the precise geographic position of Mecca. To this day, we continue to use Arabic astronomical terms such as nadir, zenith, and azimuth, and Arabic names for the stars, including Altair, Deneb, Algol, Aldebaran, Betelgeuse, Rigel, and Vega.

The Islamic world of the Middle Ages also produced two great astronomers, Abdullah Muhammed al-Battani, and Abu-al-rayhan Mohammed ibn-Ahmed al-Biruni, both of whom influenced European thought. Al-Battani, known as Albatenius in Europe, was born in Turkey in 858 A.D. and died in 929. Born a star-worshipping Sabean, he later converted to Islam.

In his writings, including his greatest work *al-Zij al-Sabi (An Astronomical Work),* Albatenius accurately calculated the length of the year and the seasons; improved Ptolemy's astronomical calculations of the planetary orbits; determined

Art by Alice Faith Dole

the true, mean orbit of the Sun; proved that (in contrast to Ptolemy's view) annular eclipses are possible; and showed that the position of the Sun's apogee is variable. Albatenius' work was so respected in Europe that Copernicus quoted him in his book, *On the Revolutions of the Heavenly Spheres* and the Danish astronomer Tycho Brahe kept a portrait of Albatenius in his observatory, next to one of Copernicus.

Al-Biruni, another astronomer, was born in Khiva, Central Asia (now U.S.S.R.) in 973, and died in 1048. Like many Moslem scholars, he made contributions to a number of scientific areas. In his two most important astronomical works, *al-Tafhim* (*Elements of Astrology*) and *Qanun al-Mas'udi* (*The Mas'udi Canon*), Al-Biruni discussed the theory of the Earth's rotation on its axis; used the stars to calculate longitudes and latitudes, and criticized the Ptolemaic model of the universe. Al-Biruni also has the distinction of being the first known writer to identify certain geologic facts, such as the formation of sedimentary rock.

Physics

Abu-Ali al-Hasan ibn al-Haytham (Alhazen in the West) was born in Baghdad in 965 and died in Cairo in 1039. Alhazen wrote more than a hundred works on math, astronomy, philosophy, and medicine.

In his main work on optics, *Kitab al-Manazir* (*Book of Optics*), he became the first to claim authoritatively that the eye receives images, refuting the work of Euclid and Ptolemy, who proposed that the eye sends out rays to perceive objects. Alhazen also projected the image of a solar eclipse into a darkened room in order to study its details, thus anticipating by centuries the "camera obscura." His influence was recognized by da Vinci, Kepler, and Roger Bacon.

Mathematics

Students in Europe and North America are taught the "Arabic numbers," algebra, and trigonometry, which are Arabic inventions. In addition, our concept of "zero" was a gift from India via the Arab empire of the Middle Ages.

The mathematician and astronomer, Muhammad ibn Musa al-Khwarizmi was born in Khiva around 780, and died in 850. Al-Khwarizmi was responsible for introducing Europe to Hindi-Arabic numerals and to algebra.

The word algebra evolved from the Arabic word, "al-jabr," in the title of his work, *Kitab al-Jabr wal Mugabla* (*The Book of Integration and Equation*), which presents examples and rules for arithmetical solutions of linear and quadratic equations, for elementary geometry, and for solving problems of proportion. This work was translated into Latin, and served as the principal math text used in European universities until the sixteenth century.

Chemistry

The term chemistry has its origin in the Arabic word "al-kimiya," or "alchemy," which is the science that so enthralled Medieval Europe and the Islamic empire, whereby attempts were made to turn base metals into gold or to discover an elixir for eternal life. These efforts were futile. However, Arab contributions to the field of chemistry were many, and words such as alcohol, alkali, antimony, soda, syrup, and alembic, which are Arabic in origin, attest to this fact.

Abu Musa Jabir ibn Hayyam is considered to be the "father of Arabic chemistry." Known as Geber in the West, he was born in Iraq in 721 and died in 815. Geber

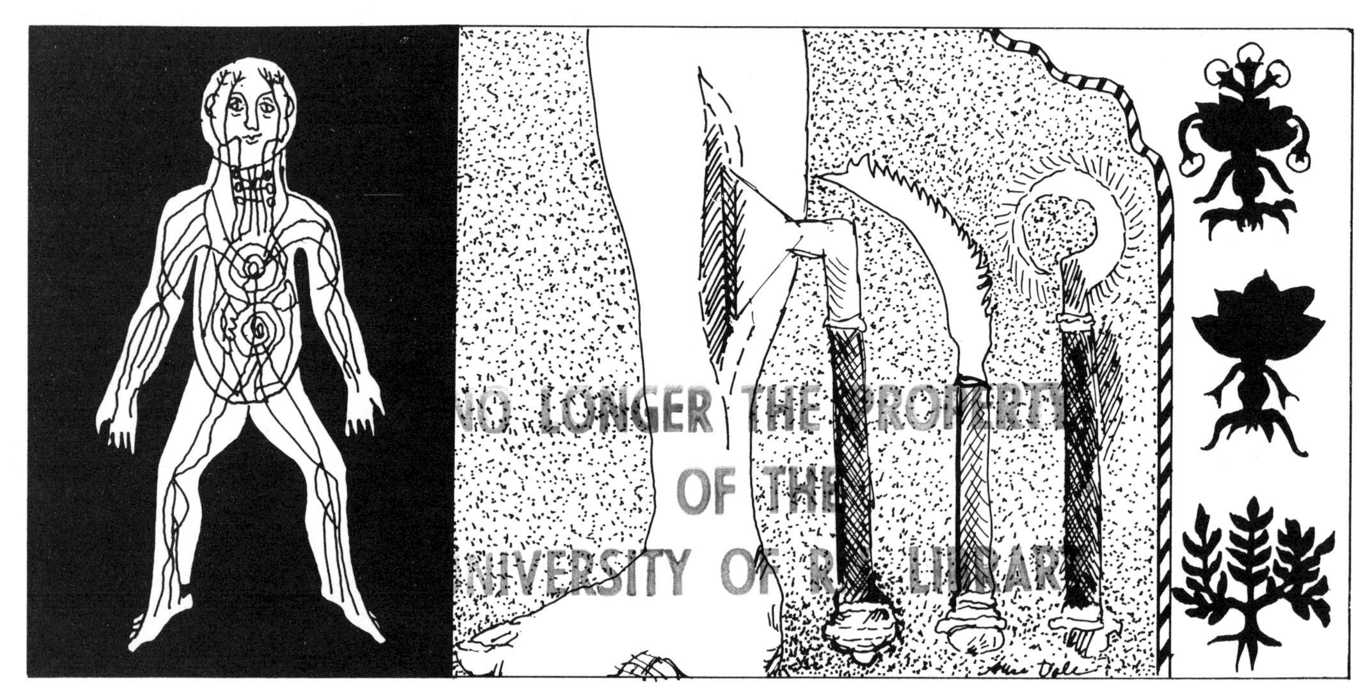

revised the ancient Greek belief that everything was made of earth, wind, fire, and water. He believed that these four "elements," when combined properly, formed mercury and sulfur, from which all other metals, including gold, were formed. He is credited with discovering several chemical compounds, scientifically describing calcination and reduction, and working on methods for evaporation, sublimation, melting, and crystallization. Geber also recognized the importance of experimentation, thereby advancing the theory and practice of chemistry.

Medicine

While European doctors were steeped in superstition and archaic methods, Islamic physicians not only enriched the classical heritage, but also advocated the study of herbs and minerals for medicinal purposes; established the first apothecary stores; advanced surgical techniques; understood the concept of contagion; and developed hospitals which were divided into wards according to disease.

The first great Moslem doctor (and one of the most prolific medical authors) was Abu-Bakr Mohammed ibn Zakariya ar-Razi, known as Rhazes in Europe. Rhazes was born in Iran in 865 and died in 923. He was a well-known alchemist before he became acquainted with medicine. He later became chief physician at hospitals in Rayy and in Baghdad.

His major medical works were *Kitab al-Mansuri* and *Kitab al-Hawi* (*The Comprehensive Book*). The *al-Mansuri* discussed a variety of medical issues, and it contains the first mention of "spiritual cure," a predecessor of modern day psychotherapy. His *al-Hawi* (unfinished at the time of his death and assembled posthumously by his students) is an encyclopedia of medical knowledge summarizing Arab, Greek, Persian, and Hindu medicine up to that time.

The greatest Arabic medieval writer on medicine was Abu Ali al Husayn ibn Sina, who was born near Bukhara, Central Asia (now U.S.S.R.) in 980 and died in 1037. Avicenna, as he came to be known in Europe, had as a young man cured the sultan of Bukhara and was given access to the royal library as a reward. By the age of 21, Avicenna had mastered the contents of the library and was ready to begin his life's work of over two hundred books on medicine, astronomy, philosophy, geometry, and theology.

His two greatest works were *Kitab al-Shifa* (*Book of Healing*), a philosophical encyclopedia based on Aristotelian, Neoplatonic and Moslem beliefs, and *al Qanun fi al-Tibb* (*The Canon of Medicine*), which was an enormous, one million word compendium based on Greco-Arab medical thought. This work replaced works by Galen and Rhazes, and became the primary medical text of European universities until the seventeenth century.

There are many others who added to our scientific knowledge, any of whom would be an excellent candidate for student research. It is important to teach students that the pursuit of scientific advancement is not a European or North American phenomenon. Many, if not all, cultures have given us insights and "giants" from whose shoulders we can see further than any previous generation. ◾

1. Simons, Grace, "Win a Nobel Prize" *Science Scope*, October, 1988.

Bruce Reichert teaches science at the American Cooperative School of Tunis, Tunis, Tunisia.